EARTH

The Making, Shaping and Workings of a Planet

EARTH

The Making, Shaping and Workings of a Planet

Derek Elsom

MACMILLAN PUBLISHING COMPANY
NEW YORK

MAXWELL MACMILLAN INTERNATIONAL
NEW YORK   OXFORD   SINGAPORE   SYDNEY

Contents

A Marshall Edition
This book was conceived, edited and designed by
Marshall Editions, 170 Piccadilly, London W1V 9DD

Macmillan Publishing Company
866 Third Avenue
New York, NY 10022

Macmillan Publishing Company is part of
the Maxwell Communication Group of Companies.

Library of Congress Cataloging-in-Publication Data

Elsom, Derek M.
     Earth : the making, shaping and workings of a planet/
Derek Elsom — 1st American ed.
         p.   cm.
     Includes index.
     ISBN 0-02-535391-8
     1. Earth.   2. Geophysics.   I. Title.
QB631.E52 1992
550—dc20   92–11079 CIP

Macmillan books are available at special discounts for bulk
purchases for sales promotions, premiums, fund-raising, or
educational use. For details, contact:
              Special Sales Director
              Macmillan Publishing Company
              866 Third Avenue
              New York, NY 10022

First American Edition 1992

Typeset by Dorchester Typesetting Group, Dorset, UK
Originated by CLG, Verona, Italy
Printed and bound in Spain by Printer Barcelona

10 9 8 7 6 5 4 3 2 1

PLANET EARTH
6

THE INTERIOR
28

SHAPING THE LAND
74

BEYOND THE LAND
116

HOW LONG CAN IT LAST?
170

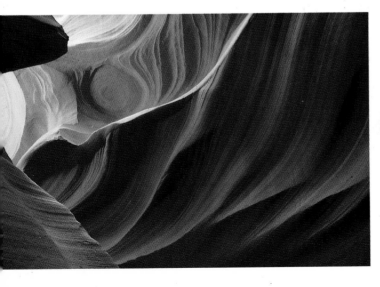

PLANET
EARTH

Planet Earth

Earth is the living planet and unique in our solar system – being the only planet with intelligent life. Whether the emergence of life was a freak accident which happened only on Earth and nowhere else in the entire universe is a question which cannot yet be answered. However, the chances of life flourishing on other planets are boosted by the knowledge that there are a billion stars in our galaxy – each with many orbiting planets – and 10,000 million galaxies in the universe.

Life and life-support systems on Earth are sustained by heat and light from the Sun, even though the amount intercepted by the planet is only 0.002 percent of the Sun's total output. That life exists on Earth is partly due to Earth's distance from the Sun. Planets adjust their surface temperature to ensure a balance in the amounts of energy received from the Sun, absorbed by the planet and reradiated back into outer space. In this way a planet neither gains nor loses heat in the long term.

Since the amount of heat received from the Sun diminishes with distance from it, the farther away a planet orbits from the Sun, the lower the surface temperature needed to achieve its heat balance. As a result, Mercury is intensely hot, while the more distant Neptune is extremely cold.

Scientists recognize a limited "life zone", or ecosphere, around a star in which life can develop if other key conditions, such as a protective atmosphere, are fulfilled. This zone is where the surface temperature of planets lies within the range for water to circulate in all of its three states – solid, liquid and vapour. Mars with its supercooled ice marks the outer edge of the ecosphere and Venus with its superheated steam lies at the inner edge. Earth the "water planet"– with oceans permanently covering some 70 percent of its surface – is in between, thus allowing the vital water cycle to operate.

The planet's atmosphere plays a role, too, in determining whether its surface favours life. Greenhouse gases – water vapour, carbon dioxide and methane – trap outgoing heat and act like an insulating blanket, raising surface temperature above the value determined simply by distance from the Sun. Although the Moon is a similar distance from the Sun, it lacks an atmosphere so the average surface temperature is a freezing 3°F (−16°C), compared with Earth's average of 59°F (15°C).

Neither are the atmospheres of the other planets in our solar system conducive to life. Mars's is so thin that its greenhouse gases make little difference to its surface temperature of about −108°F (−77°C). Venus's dense atmosphere is saturated with carbon dioxide forced out of its carbonate rocks by intense solar heat. Levels of carbon dioxide reach 96 percent – compared with Earth's 0.04 percent – and a surface temperature of 896°F (480°C) results.

The mix of gases in the Earth's atmosphere is strikingly different from other planets, with large amounts of oxygen and nitrogen. This differs radically from what would be expected from the physical processes of

Sudden squally storms rush up from the Antarctic, bringing ice-laden winds and heavy rain (previous page). Here a hailstorm approaches the western tip of Campbell Island, south of New Zealand. Offshore are the jagged, weathered rocks of Dent Island.

planet develop-ment. It seems that life has modified the Earth's initial atmosphere to suit its own survival. Biological organisms extracted large amounts of carbon dioxide from the early atmosphere through photosynthesis for use by plants, phytoplankton and blue-green algae while releasing oxygen for animal consumption.

To maintain optimum warmth, our living planet manipulates not only the composition of the atmosphere but also the Earth's surface. Changing the area covered by clouds or vegetation alters the amount of solar energy reflected. In this way life-support systems behave like global thermostats. If temperatures rise, more evaporation generates more cloud cover which reduces the amount of sunlight received by the surface, so restoring the temperature balance.

Computer simulations of the creation of the solar system, showing the birth of the planets by cataclysmic collisions between hundreds of swirling chunks of debris, suggest that the Earth's position relative to the Sun in the ecosphere is not purely random. These images reveal that a planet similar to Earth in size usually forms at about its present orbital distance from the Sun. Such a result suggests where to seek out life around other stars.

Distance from the Sun may initially have been a critical factor for life to begin — although chemical reactions were still vital to create the carbon compounds from which bacteria could eventually evolve. From that point on, it appears that life on Earth has influenced its own destiny by changing the environment to suit itself. Its purpose seems to be to increase the complexity, intelligence and diversity of biological organisms.

A great variety of plant and animal species is encouraged by the existence of seasons caused by the angle of tilt of the planet's axis. This remains fixed in space as the Earth orbits the Sun, so that each hemisphere tilts toward and then away from the Sun, creating warm summers and cold winters in middle and high latitudes.

Interaction between life and the environment has resulted in the evolution of a number of complex and powerful cycles including those of water, carbon, nitrogen, oxygen, phosphorus and sulphur. Cycles of life and death ensure energy and nutrients are passed along food chains as soil microorganisms, bacteria and fungi regenerate nutrients after the death of a plant or animal. These self-regulating cycles interrelate to maintain suitable conditions for the living planet to be true to its name.

Life-support systems are now under threat. Human activities are polluting the environment, eroding soils, destroying habitats, depleting natural resources and causing the extinction of a great many plant and animal species. For too long we have acted without due consideration of the consequences of our actions. It is now time for us to recognize our responsibility toward safeguarding the future health of Earth. Attempting to increase our understanding of the complex workings and intricate systems which sustain the living engine that is our planet is a beginning.

Cosmic connections

Nebula

Protostar

Neil Armstrong, the first person to walk on the Moon, described the Earth with its deep blue oceans, brown-green continents and swirling veil of dazzling white clouds as "a beautiful jewel in space". But probably what makes it unique is that intelligent life evolved on this small planet and, as far as we know, only on this planet out of trillions across the universe.

The young Earth provided all the conditions for carbon-based life to evolve: a protective atmosphere; organic chemicals; circulating elements, including water in its three states; volcanic activity to link crust and atmosphere; and a surface temperature at which the appropriate chemical reactions could take place.

Planet Earth seems an insignificant body in a vast universe, which was created, according to the big bang theory, when a concentration of energy and matter exploded apart 15,000 to 20,000 million years ago. Observations of the most distant stars show it is still expanding.

Around 4,600 million years ago one of many swirling clouds of gas, or nebulae, in the universe formed a star – the Sun – at its centre. The surrounding dust grains joined together to create nine orbiting planets and smaller bodies such as moons, asteroids and comets. The Earth's moon is unusually large and is probably an embryonic planet which failed to grow and compete as a planet in its own right.

The fast-moving inner planets are Mercury, Venus, Earth and Mars. They have relatively thin atmospheres, since the Sun's heat boiled away lighter materials. The larger outer

Planet	1 Mercury	2 Venus	3 Earth	4 Mars
Diameter	3,030 mi (4,880 km)	7,520 mi (12,100 km)	7,930 mi (12,760 km)	4,220 mi (6,790 km)
Distance from Sun	36 million mi (58 million km)	67 million mi (108 million km)	93 million mi (150 million km)	142 million mi (228 million km)
Number of moons	0	0	1	2
Day (rotation period)	58.6 days	243 days	24 hours	24.6 hours
Year (time to orbit Sun)	88 days	225 days	365 days	687 days

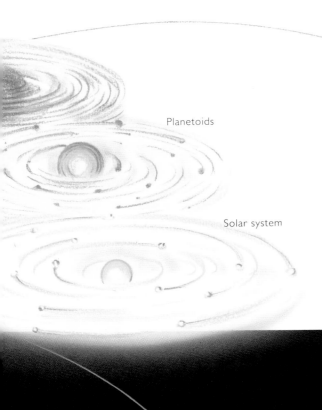

Planetoids

Solar system

Some 4,560 million years ago, the solar system was created when a cold swirling gas cloud, or nebula, near the edge of the spiral Milky Way galaxy caved in. Its centre contracted and increased in density causing it to heat up.

This triggered thermonuclear fusion, creating a protostar radiating heat and light. Dust grains formed from the condensing gases and collided, fusing eventually into large chunks. High-speed collisions between these planetoids, up to 600 miles (1,000 km) across, produced the nine planets.

planets, Jupiter, Saturn, Uranus and Neptune, retained the lighter gases and have solid cores surrounded by vast cold atmospheres of methane, ammonia, helium and hydrogen. Pluto, the tiny outermost planet, has a highly eccentric and steeply inclined orbit, which means it is sometimes closer to the Sun than Neptune.

The immense size of the universe, with its 10,000 million galaxies, is not easy to grasp. Light travelling from the Sun to the Earth at 186,000 miles per second (300,000 km/sec) takes 8.3 minutes to travel the 93 million miles (150 million km). By contrast, light from the Moon takes only a second to reach the Earth. At the other end of the scale, to traverse our galaxy would take 100,000 years travelling at the speed of light.

The solar system contains nine planets, some with orbiting moons, revolving in near-circular orbits around a slowly rotating Sun.

Small rocky planets lie close to the Sun – Mercury, Venus, Earth and Mars – while giant gaseous ones lie farther out, with a small icy planet, Pluto, and its moon Charon, at the outer edge. Some outer planets collected their own moons and resemble miniature solar systems.

Signs of a failed attempt at planet formation are evident in the asteroid belt between Mars and Jupiter. Other bodies in the solar system include billions of lumps of ice a few miles across forming comets beyond the orbit of Pluto.

Because of unexplained orbital variations in the outer planets, some astronomers have believed that a tenth far-flung planet exists. However, many searches have failed to find it and the odds on its existence are dwindling.

	5 Jupiter	6 Saturn	7 Uranus	8 Neptune	9 Pluto
	88,850 mi (142,980 km)	74,900 mi (120,540 km)	31,765 mi (51,120 km)	30,780 mi (49,530 km)	1,430 mi (2,300 km)
	483 million mi (778 million km)	890 million mi (1,430 million km)	1,780 million mi (2,870 million km)	2,795 million mi (4,500 million km)	3,675 million mi (5,915 million km)
	16	18	15	8	1
	9.8 hours	10.2 hours	17.9 hours	19.2 hours	6.4 days
	11.9 years	29.5 years	84 years	165 years	248.5 years

Planetary forces

Invisible curving lines of magnetic force arch out into space creating a shield which affords the Earth protection from the lethal bombardment of high-energy particles, which stream from the Sun at a million miles an hour (1.6 million km/h). This magnetic shield deflects the particles around it and creates a protective cocoon, called the magnetosphere, surrounding the Earth.

The force field is not totally effective so some solar wind particles are able to reach the upper atmosphere, or ionosphere, around the magnetic poles. There they strike oxygen and nitrogen atoms and molecules, exciting them to emit light, creating the spectacular shimmering visible in the night sky as the aurora borealis, or northern lights, and aurora australis, or southern lights. Aurorae occur in an oval zone between 60 and 75 degrees of latitude around the magnetic pole but after a strong solar flare the increased solar wind pressure pushes them slightly equatorward.

The distortion of the magnetic field and formation of intense aurorae following a solar flare can disrupt radio communications as well as induce power surges in electrical transmission lines. In March 1989 circuit breakers triggered, leaving millions in Canada and Sweden without power. In June 1991 communication satellites which rely on the Earth's magnetic field for orientation became confused and turned upside down. Ground operators corrected the problem using tiny on-board rockets.

The lines of magnetic force flow downward at the magnetic north pole and upward at the south pole. When volcanic lava solidifies, its small iron crystals behave like compasses and lock in the magnetic dip and direction of the magnetic north pole at that time.

Magnetic analyses of ancient lava samples reveal that the continents have drifted across the planet's surface. Rock samples also reveal that the direction of the Earth's magnetic field has reversed every few hundred thousand years. When magnetic reversals occur there may be an interval when the magnetic field is virtually absent, causing severe problems for life on Earth especially for birds relying on the Earth's magnetic field for seasonal migrations.

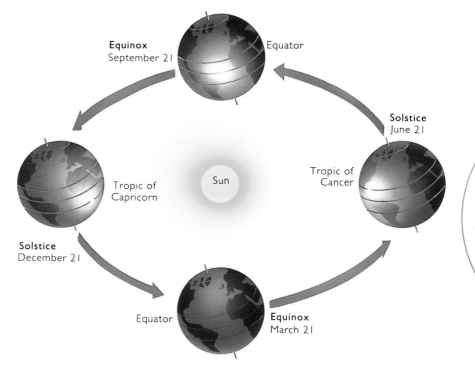

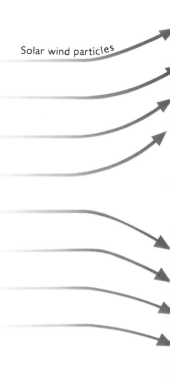

Solar wind particles

The cycle of seasons occurs because the axis of the spinning Earth is tilted at 23.5° from the perpendicular to an imaginary line joining the centres of the Sun and the Earth. As the Earth revolves around the Sun, the axis alignment remains fixed in space so that first one hemisphere and then the other leans toward the Sun, producing summer in each hemisphere.

Seen from the northern hemisphere, the Sun's midday position climbs higher in the sky as summer arrives. The longest day is on June 21, the summer solstice, when the Sun lies over the Tropic of Cancer. In winter the Sun shifts into the southern hemisphere creating the shortest day on December 21, the winter solstice, when the Sun is over the Tropic of Capricorn.

When the Sun lies above the equator, the spring and autumn equinoxes, day and night are equal in length.

The Earth's magnetic field is like a giant teardrop extending far out into space. The shape is that which would be formed if the core of the planet held a powerful bar magnet or a metal coil with an electrical current flowing through it.

Most geophysicists agree that the magnetic field is generated by eddies. These are driven by heat released by radioactive elements in the electrically conductive iron-rich liquid of the Earth's outer core. Where supersonic solar wind particles strike this force field they slow down producing a bow shock.

Inside this lies the magnetopause, which is the boundary of the magnetosphere. Van Allen radiation belts trap particles which penetrate the magnetopause but some are funnelled down the open field lines of the polar clefts where they strike atoms and molecules in the ionosphere producing aurorae.

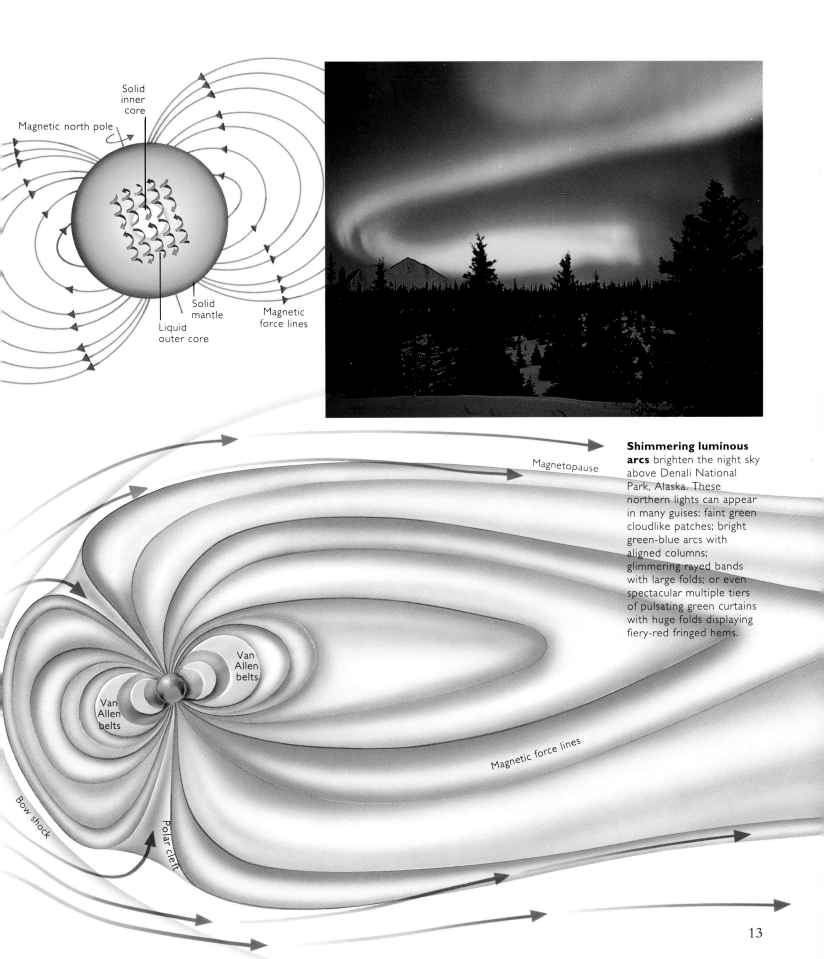

Solid
inner
core

Magnetic north pole

Solid
mantle

Liquid
outer core

Magnetic
force lines

Shimmering luminous arcs brighten the night sky above Denali National Park, Alaska. These northern lights can appear in many guises: faint green cloudlike patches; bright green-blue arcs with aligned columns; glimmering rayed bands with large folds; or even spectacular multiple tiers of pulsating green curtains with huge folds displaying fiery-red fringed hems.

Magnetopause

Van
Allen
belts

Van
Allen
belts

Bow shock

Polar cleft

Magnetic force lines

13

The emergence of life

Soon after planet Earth was created, the cosmic gases which formed its initial thick atmosphere were stripped away by a solar wind of fast-moving particles flung out from the newborn Sun. When volatile gases escaped from the planet's hot interior through the outer crust as the Earth cooled and solidified, the atmosphere began to re-form. Volcanic eruptions added further gases until a relatively thin atmosphere built up from oxidized carbon, nitrogen and hydrogen gases.

There was no free oxygen until 2,000 million years ago. Only then did primitive sea-dwelling plants and algae begin releasing oxygen and removing carbon dioxide at the same time. More oxygen meant that complex life forms could evolve. By their photosynthesis, green plants increased oxygen levels until eventually 21 percent of the atmosphere was oxygen.

Early life forms had to be protected from the Sun's ultraviolet radiation which damages living cells. Today, the stratospheric ozone layer absorbs most of these incoming rays and allows organisms to thrive. But until oxygen levels in the atmosphere were sufficient to generate ozone (a molecule containing three oxygen atoms) life had an alternative filter: ocean water, about 33 feet (10 m) deep, filtered out ultraviolet yet allowed enough light to reach organisms for photosynthesis. Only as atmospheric oxygen increased and ozone levels rose could life inhabit the ocean surface or move on to the exposed land.

Species have evolved, thrived, dominated, diminished and become extinct throughout the planet's history. The average time span for a species is only about five million years, although some have lasted considerably longer. As species have died they have left vacant niches to be filled. Equally, new niches have been created as species have exploited new environments produced by climate changes and the spreading apart of continents.

At many times in the Earth's history there have been large-scale species extinctions. The most likely cause of the extinction of the dinosaurs, ammonites and many other species around 65 million years ago was that one or more asteroids collided with the Earth.

In an expanding time spiral Earth's evolution is traced since its surface began to solidify 4,000 million years ago. As this crust cracked, re-formed and thickened, volcanic eruptions poured out vast amounts of red-hot gases.

As the surface cooled, water vapour condensed. Rain was made, which filled the basins to form the oceans. Chemical reactions triggered by the Sun's ultraviolet rays, lightning or the shock waves from a meteor strike concocted various carbon compounds including amino acids which provide the building blocks for life. Fossilized evidence of life 3,500 million years ago comes from Australia where rock mounds, or stromatolites, were formed by bands of primitive ocean algae.

A critical point came 3,000 million years ago when blue-green algae began to photosynthesize and liberate oxygen. As oxygen levels increased ocean animals arose, for example jellyfish some 600 million years ago. By 570 million years ago the first animals with shells evolved, and then fish with backbones. Plants and animals moved on to the land some 400 million years ago.

Tropical forests and swamps expanded to be populated by worms, spiders and insects and later by reptiles, amphibians and winged insects. One group of reptiles evolved into dinosaurs, another into early mammals around 225 million years ago. While dinosaurs ruled the world, mammals remained nocturnal creatures but 65 million years ago mammals took over and the ancestors of cows, horses, elephants and finally humans arrived.

Conifers

Giant dragonfl

Reptiles *Hylonomus*

Giant ferns
Tree fern

Giant horsetail

Fish *Platysomus*

Cephalopods

Dimetrodon

Mayfly

Crinoids

Mammal-like reptiles

Echinoids

Euparkeria

Early dinosaurs *Coelophysis*

Teleost fish *Pholidophorus*

Gingko

Ichthyosaurs

Triassic

Jura

Early pterosaurs

Apatosaurus

Ammonites

Tortoise

Cycad

Belemnites

Benetiales

Stegosaurus

Modern fish *Thrissops*

First birds *Archaeopteryx*

Sauropods *Brachiosaurus*

■ Mass extinction

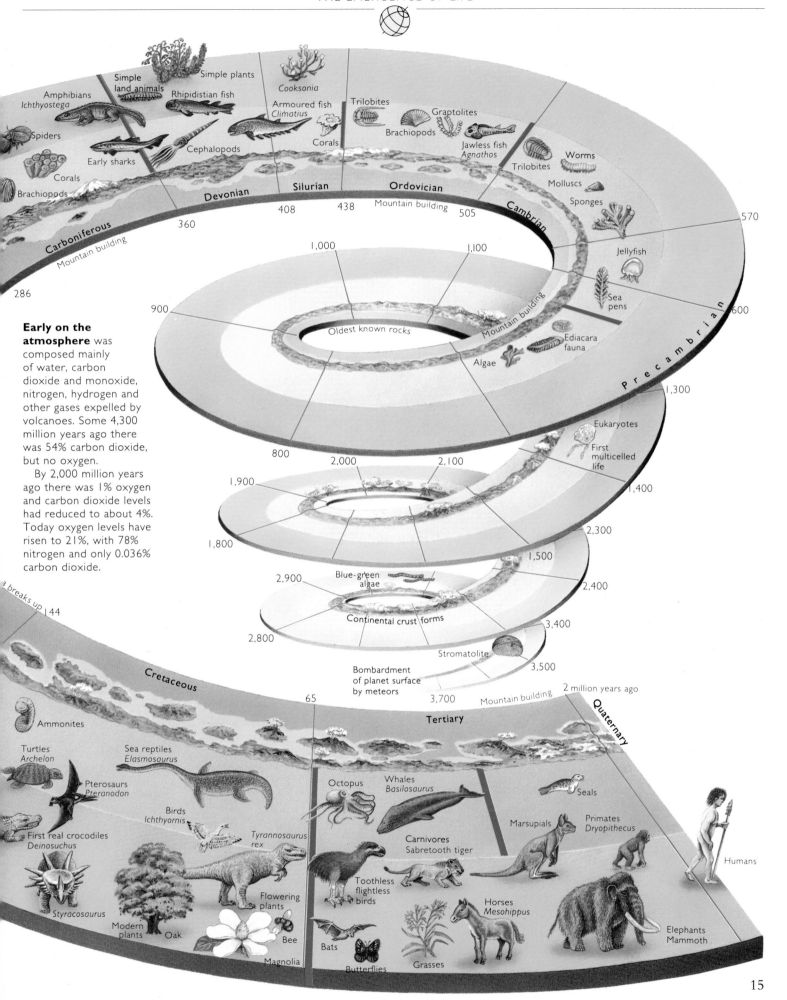

Simple plants

Simple land animals

Amphibians
Ichthyostega

Rhipidistian fish

Cooksonia

Armoured fish
Climatius

Trilobites

Graptolites

Brachiopods

Spiders

Cephalopods

Early sharks

Corals

Corals

Jawless fish
Agnathos

Trilobites

Worms

Molluscs

Sponges

Brachiopods

Devonian

Silurian

Ordovician

Mountain building

Cambrian

Carboniferous

360

408

438

505

570

Mountain building

286

1,000

1,100

Jellyfish

900

Oldest known rocks

Mountain building

Sea pens

600

Early on the atmosphere was composed mainly of water, carbon dioxide and monoxide, nitrogen, hydrogen and other gases expelled by volcanoes. Some 4,300 million years ago there was 54% carbon dioxide, but no oxygen.

By 2,000 million years ago there was 1% oxygen and carbon dioxide levels had reduced to about 4%. Today oxygen levels have risen to 21%, with 78% nitrogen and only 0.036% carbon dioxide.

Ediacara fauna

Algae

1,300

800

2,000

2,100

Eukaryotes

First multicelled life

1,900

1,400

1,800

2,300

1,500

2,900

Blue-green algae

2,400

Continental crust forms

2,800

3,400

a breaks up 144

Stromatolite

3,500

Cretaceous

65

Bombardment of planet surface by meteors

3,700

Mountain building

2 million years ago

Quaternary

Ammonites

Tertiary

Turtles
Archelon

Sea reptiles
Elasmosaurus

Octopus

Whales
Basilosaurus

Seals

Pterosaurs
Pteranodon

Birds
Ichthyornis

Tyrannosaurus rex

Marsupials

Primates
Dryopithecus

First real crocodiles
Deinosuchus

Carnivores
Sabretooth tiger

Humans

Styracosaurus

Flowering plants

Toothless flightless birds

Horses
Mesohippus

Modern plants

Oak

Bee

Bats

Grasses

Elephants
Mammoth

Magnolia

Butterflies

P r e c a m b r i a n

15

The emergence of life/2

630 Precambrian

505 Cambrian

438 Ordovician

360 Silurian

Devonian

Carboniferous

248 Permian

213 Triassic

Jurassic

Million
years
ago   144

Mass extinction

Cretaceous

Alamosaurus

Hypacrosaurus

Parasaurolophus

Pteranodon

Tyrannosaurus rex

Ornithomimus

Pachycephalosaurus

Archelon

Ankylosaurus

Triceratops

Ischyodus

Hesperornis

Champsosaurus

Purgatorius

Ammonites

Elasmosaurus

Sphenocephalus

Rudist
bivalves

Enchodus

Mass extinction

65 million years ago

Survival on Earth is not guaranteed for any species – as the dinosaurs found out to their cost 65 million years ago. At that time nearly three-quarters of all plant and animal species, including the reptiles that had dominated the planet for tens of million years, became extinct. Geologically, this massive change is marked by the KT boundary, when the Cretaceous period gave way to the Tertiary. (K comes from *Kreide*, the German for Cretaceous.)

One species' loss is another's gain. Small, often nocturnal, mammals – through luck or inborn tolerance to harsh environmental conditions – survived. They quickly exploited newly vacant ecological niches and soon came to dominate the planet as the reptiles had before them. It is from these mammals that humanlike species evolved around 5–10 million years ago.

The cause of mass extinctions is hotly debated. Prolonged cold and darkness caused by dust and gases sent high into the atmosphere by an asteroid striking the planet and/or from a massive volcanic eruption are the most likely reasons. Plants could not survive without photosynthesizing and this would have caused a collapse in the food chain triggering a cascading ecological disaster as animals starved to death.

The mass extinction around 65 million years ago was one of many that the Earth has suffered. Around 248 million years ago an astounding 96 percent of all species disappeared. There may be a cycle of extinctions but fortunately the next one is not due for 15 million years or so.

The extensive dust plumes ejected into the air from the vaporized remains of the asteroid and surface rock enveloped the planet, blocking out sunlight and creating years – even hundreds of years – of prolonged darkness and cold.

Evidence to support a meteor impact is a layer of clay, 65 million years old, which contains iridium, a metal that is rare on Earth but common in some meteorites, as well as deformed quartz crystals and tiny beads of glass which are known to be produced by meteor impacts. Possible strike sites include the 20-mile (32-km) Manson Crater, Iowa, and the 110-mile (177-km) Chicxulub Crater in the Yucatán. Additional climate cooling then may have come from gases ejected during a massive outpouring of lava that now forms the Deccan Traps of India.

Despite their dominance of the planet today, humans have existed for an incredibly short part of the planet's history. Hominids, our humanlike ancestors, diverged from apes 5–10 million years ago in east Africa. When a drier climate encouraged grasslands to develop, some tree-living apes left their tropical forests and entered these open areas.

They adapted gradually, prospered and became a separate species changing from being short, broad and stooped to being tall, slim and erect. In the east African cradle of human evolution *Homo erectus* developed stone tools two million years ago and then migrated toward Europe, Arabia and China. By 500,000 years ago this species could control fire, allowing it to survive the chilly winters of middle latitudes.

Modern humans, *Homo sapiens sapiens*, evolved around 90,000 years ago. By 30,000 years ago they had expanded into Australia and within a few thousand years crossed the Siberia-Alaska land bridge and spread down the west coast of North America. Around that time the human race consisted of only five million fur-clad hunter-gatherers. Today, a human population explosion has produced nearly six billion humans who threaten to cause another episode of species extinction as the planet's natural ecosystems are destroyed, altered and polluted.

Birds

Planetetherium

Miacis

Tortoises

Procoptodon

Snakes

Stylinodon

Phorusrhacus

Palaeoryctes

Purgatorius

Pristichampsus

Champsosaurus

Tertiary

Ischyodus

Turtles

Hypsidoris

Enchodus

Earth's heat engine

Living things – as well as the rocks, air and oceans of our planet – are woven together into a single entity by flows of energy, water and nutrients within and around the planet. This living engine that drives planet Earth is called Gaia, the name given by the ancient Greeks to their Earth goddess.

The notion of Gaia is that life on Earth owes its existence to its intrinsic ability to manipulate global climate and thus maintain the cycles of chemical elements essential for its survival. Its primary control is achieved through its influence on the planet's brightness, or albedo, and on the gas content of the atmosphere, which is strikingly different from other planets in the solar system. Together these act as global temperature regulators, keeping the climate of planet Earth within life-sustaining limits.

A mean, or average, global surface temperature of around 55°F (13°C) has been maintained since life first evolved more than 2,000 million years ago. Even during the ice ages, periods forced upon the planet largely by a wobble in the tilt of the Earth of two degrees from normal, the tropics were for the most part unaffected, and mean global temperature dropped by only 9°F (5°C).

The result of this regulation is that precious planetary water neither freezes over entirely nor dries completely, and that life can flourish. As the cycle of plant life continues, so it controls global temperature as vegetation expands or contracts in area. Where vegetation thrives, making the surface appear dark in colour, sunlight is absorbed, warming the planet's surface. Conversely, where vegetation is absent or pale in colour, light is reflected back through the atmosphere and returned to space. This causes a cooling of the surface.

At times of rising temperatures, the millions of tiny plants that form the oceans' phytoplankton cool the planet. They do this by releasing, as a result of their metabolic processes, the chemical dimethyl sulphide, which is an effective cloud-forming agent. The resultant increase in cloud cover blocks out some of the Sun's energy and the Earth's temperature drops.

The Earth's surface temperature is maintained by a kind of "heat engine" which controls the flow of energy. In this diagram, the amounts of incoming and outgoing solar radiation in the engine are drawn to scale to show the relative strengths of energy flow as a percentage of the total. The amount of radiation from the Sun (100%) equals the amount leaving the atmosphere, some of which is reflected (26% + 4%) and some of which is radiated (65% + 5%).

Water vapour, carbon dioxide and other trace gases in the atmosphere absorb some of the Earth's outgoing radiation and reradiate it back to the ground. This means that between the atmosphere and the Earth 132 percent of energy is circulating at any one time. This provides extra warmth, like an insulating blanket, which maintains a comfortable average surface temperature – the natural greenhouse effect.

Human reactions are increasing the amounts of atmospheric greenhouse gases, disrupting the planet's finely tuned heat engine mechanisms and forcing temperatures to rise to extremes never experienced before. Average global temperature has increased by 0.9°F (0.5°C) in the past 100 years.

Radiation from the Sun (100%)

Reflection by Earth's surface (4%)

Reflection by atmosphere (26%)

Short-wave solar radiation

Long-wave terrestrial radiation

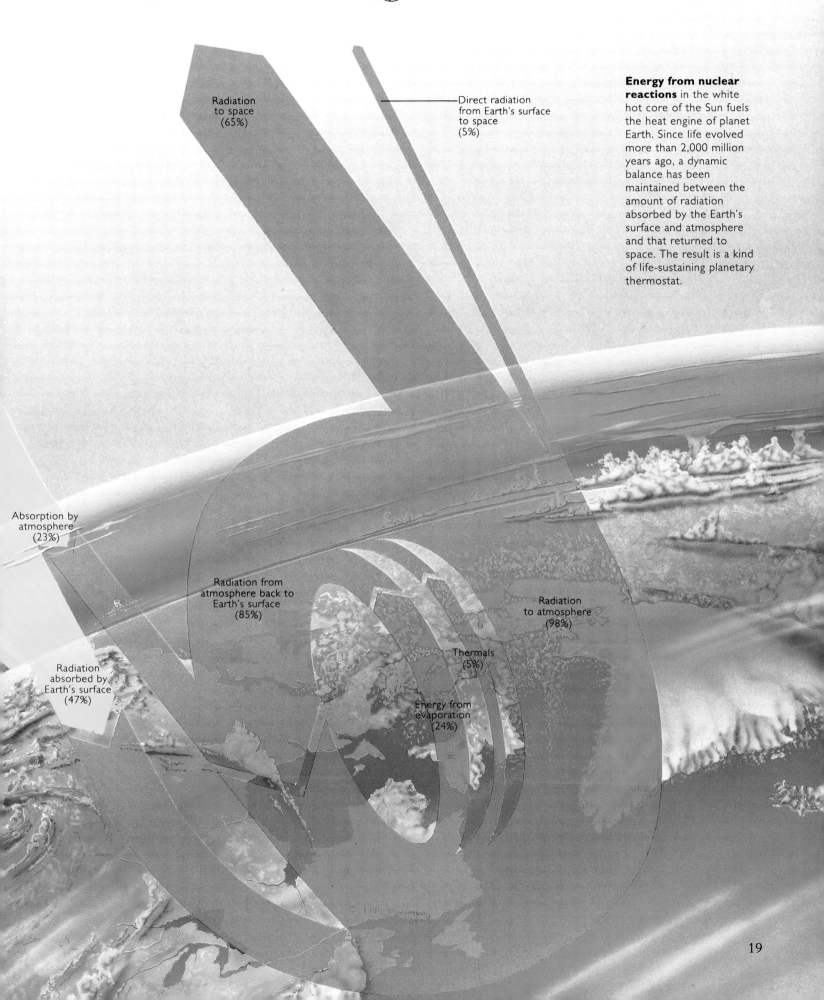

Radiation
to space
(65%)

Direct radiation
from Earth's surface
to space
(5%)

Energy from nuclear reactions in the white hot core of the Sun fuels the heat engine of planet Earth. Since life evolved more than 2,000 million years ago, a dynamic balance has been maintained between the amount of radiation absorbed by the Earth's surface and atmosphere and that returned to space. The result is a kind of life-sustaining planetary thermostat.

Absorption by
atmosphere
(23%)

Radiation from
atmosphere back to
Earth's surface
(85%)

Radiation
to atmosphere
(98%)

Thermals
(5%)

Radiation
absorbed by
Earth's surface
(47%)

Energy from
evaporation
(24%)

19

Water: life's essential molecule

In the words of Leonardo da Vinci, "water is the driver of nature". It is one of the commonest molecules circulating on Earth, with a global volume of 330 million cubic miles (1,400 million cu km). Although water is essential to life, its regional distribution varies greatly from parched deserts to deep oceans. On land the amount and timing of rain are vital. If too much falls too fast devastating floods can follow, as in Bangladesh and China. If the rains are too infrequent or the amount too small, severe drought can result, as in Ethiopia.

The water cycle operates on a local level as well as on a global scale, and can be viewed as a balancing act. The amount of water which flows out of a river basin will equal the amount of water which falls from the sky, less that lost by transpiration and evaporation in the atmosphere, together with any losses when the water percolates through the ground.

The oceans store by far the greatest amount of water, but its saltiness means people have to rely on rivers and underground water for drinking, irrigation and industrial processing. Rocks below the surface with tiny voids between their grains, such as porous sandstone, provide an underground reservoir for water. Where the pores are joined, a network of miniature water channels allows water to flow through the permeable rock. Water-laden rock is an aquifer and the level to which it is filled with water is its water table.

The Earth's water is becoming increasingly contaminated. Described as Europe's largest sewer, the River Rhine supplies drinking water for 30 million people. However, it is probably "cleaner" than drinking water in some parts of the world, where 17,000 children die from diarrhoeal diseases every day. Developed nations consume vast amounts of clean water for washing clothes, dishes and cars, yet there are more than one billion people in developing nations who have no clean water to drink.

It is not only river water that is dirty; the rains and snows of Europe and North America also carry pollutants. Chemicals discharged into the atmosphere are picked up by this precipitation and returned to the Earth's surface in a never-ending cycle.

A water molecule can take millions of years to complete a cycle or just a few hours; the average time it remains in the atmosphere is 10 days.

The Sun's energy powers the water cycle by heating the oceans, lakes and land; as a result water is evaporated into the atmosphere. Evaporation from the oceans provides the largest transfer of water vapour to the atmosphere.

Moisture-laden air is carried by winds to other parts of the ocean or to the continents, where it condenses into clouds and produces rain or snow. Over mountains and at high latitudes, snow feeds ice sheets and glaciers and can be trapped as ice for thousands of years until it melts in a warm period.

Rain either seeps into the soil or flows as surface runoff into rivers and lakes before returning to the oceans with sediments eroded by the river.

Water seeping down through soil may reach solid rock and collect in cracks. Aquifers, rocks saturated with ground-water, can store water for thousands of years or provide a steady flow to springs, rivers and wells.

In the tropics a daily cycle is in motion, with the morning sun promoting evaporation of surface water and the build up of clouds, which produce showers by the afternoon.

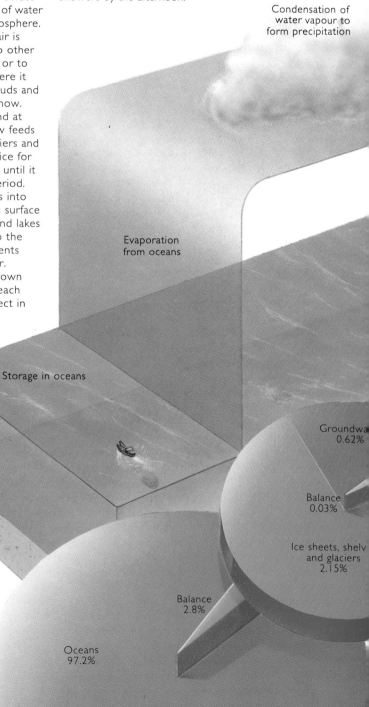

Condensation of water vapour to form precipitation

Evaporation from oceans

Storage in oceans

Groundwa 0.62%

Balance 0.03%

Ice sheets, shelv and glaciers 2.15%

Balance 2.8%

Oceans 97.2%

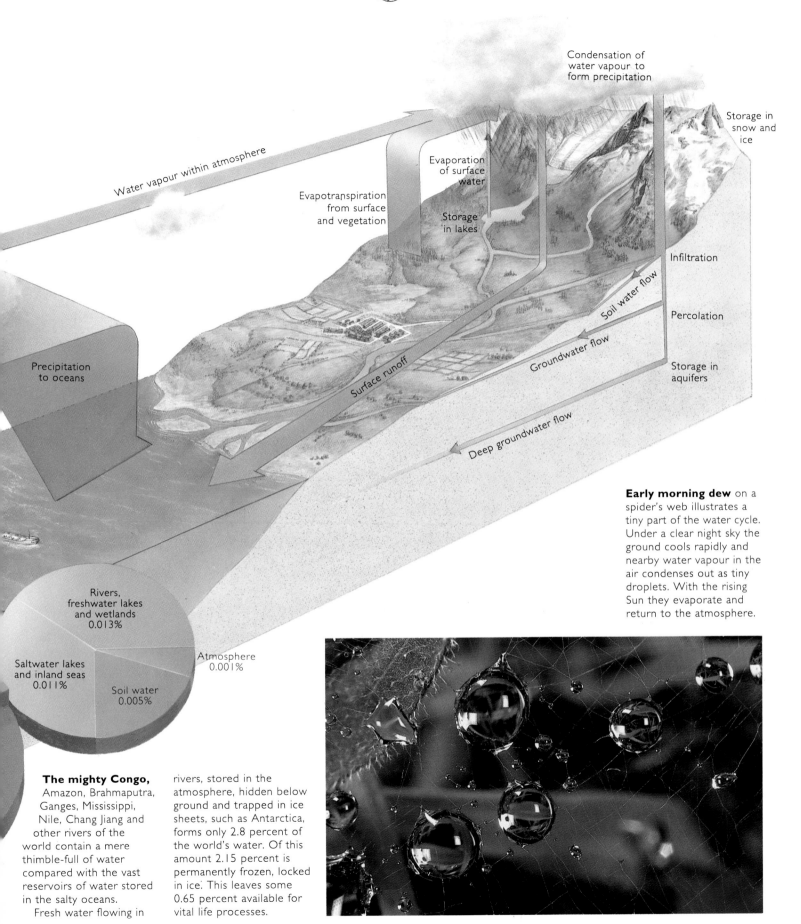

Condensation of
water vapour to
form precipitation

Storage in
snow and
ice

Water vapour within atmosphere

Evaporation
of surface
water

Evapotranspiration
from surface
and vegetation

Storage
in lakes

Infiltration

Soil water flow

Percolation

Precipitation
to oceans

Surface runoff

Groundwater flow

Storage in
aquifers

Deep groundwater flow

Rivers,
freshwater lakes
and wetlands
0.013%

Atmosphere
0.001%

Saltwater lakes
and inland seas
0.011%

Soil water
0.005%

Early morning dew on a
spider's web illustrates a
tiny part of the water cycle.
Under a clear night sky the
ground cools rapidly and
nearby water vapour in the
air condenses out as tiny
droplets. With the rising
Sun they evaporate and
return to the atmosphere.

The mighty Congo,
Amazon, Brahmaputra,
Ganges, Mississippi,
Nile, Chang Jiang and
other rivers of the
world contain a mere
thimble-full of water
compared with the vast
reservoirs of water stored
in the salty oceans.
    Fresh water flowing in
rivers, stored in the
atmosphere, hidden below
ground and trapped in ice
sheets, such as Antarctica,
forms only 2.8 percent of
the world's water. Of this
amount 2.15 percent is
permanently frozen, locked
in ice. This leaves some
0.65 percent available for
vital life processes.

Vital cycles

Carbon and oxygen, two elements found in abundance on our planet, are vital to life. Carbon provides the foundation of all plant and animal life; oxygen ensures that they can breathe. Carbon is also the basis for many compounds, from diamonds to coal, while oxygen accounts for a quarter of all atoms in living matter. In their gaseous states both elements are part of the atmosphere which protects life from harmful solar and cosmic radiation. They combine to produce carbon dioxide, a greenhouse gas, which traps terrestrial long-wave radiation and ensures life on Earth enjoys tolerable temperatures.

Plants on land and in the seas are the key to the carbon and oxygen cycles. They use the green pigment chlorophyll to trap sunlight and pass it through a chain of chemical reactions involving carbon dioxide and water in a process called photosynthesis. This converts sunlight into chemical energy in the form of sugars and other carbohydrates, which plants need to survive and grow. As a by-product of the splitting of water during photosynthesis, plants release oxygen.

Plants, eaten by animals, transfer energy and provide the basis of the food chain. The oxygen released by plants is consumed by animals, which in turn release carbon dioxide and water vapour as by-products of respiration. Every year, land plants take up 120 billion tons of atmospheric carbon − 24 times the amount

Virgin tropical rainforests, as here in Brazil, act as huge lungs storing vast amounts of carbon which they extract from the atmosphere and releasing oxygen as a by-product of photosynthesis.

Such forests are fast being destroyed in Brazil, Colombia, Indonesia, Laos and Thailand to make way for hydroelectric dams, open-cast mining, cattle ranching and roads. This must stop if we are not to suffer the consequences of disrupting the global carbon and oxygen cycles.

In a process known as photosynthesis plants use sunlight to convert carbon dioxide in the air or in sea water into the carbohydrates needed for growth. These plants form the basis of the food chain when eaten by animals.

When plants and animals die they decompose in the soil or sea and release carbon into the atmosphere. Some organic matter falls to the sea floor where, over millions of years, it forms limestone and chalk. It may be uplifted later and weathered by rainfall, which releases the carbon. Plants buried on land can turn to coal, which releases carbon when burned.

Excessive use of coal and oil is producing a rapid rise in carbon dioxide in the atmosphere, leading to global warming. Removing carbon dioxide from power station emissions could help, but the gas would need to be stored so it did not quickly rejoin the carbon cycle. Solutions include pumping it into abandoned gas fields or the oceans at places where water sinks to the sea floor and does not surface again for a thousand years.

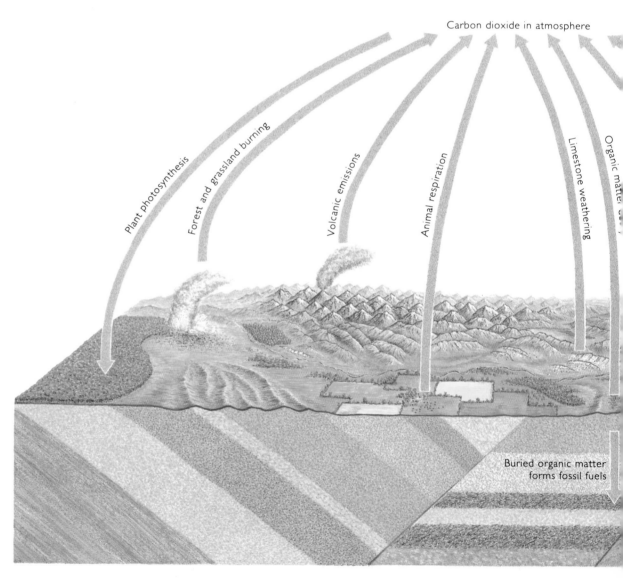

Carbon dioxide in atmosphere

Plant photosynthesis

Forest and grassland burning

Volcanic emissions

Animal respiration

Limestone weathering

Organic matter

Buried organic matter forms fossil fuels

released by fossil fuel burning – but a similar amount is returned to the atmosphere as animals breathe and dead plants and animals are decomposed by microorganisms in the soil.

Human activities are beginning to upset the global balance of the cycle. The burning of fossil fuels releases 5 billion tons of carbon dioxide into the atmosphere each year; clearing forests adds another 2 billion tons. Enhanced plant growth and increased absorption of carbon dioxide by phytoplankton in the oceans have offset only some of this increase. Atmospheric carbon dioxide levels have risen by 25 percent in the past century and global surface temperature by 0.9°F (0.5°C), raising concerns of runaway global warming.

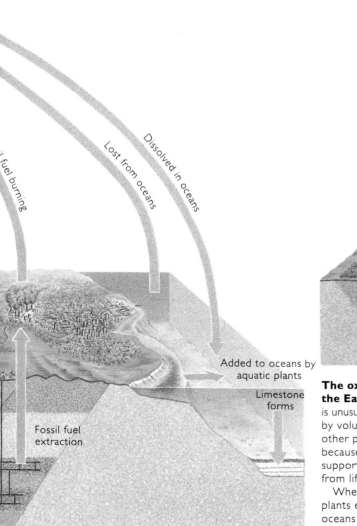

Oxygen in atmosphere

Plant photosynthesis

Plant respiration

Volcanic emissions

Oxidation of minerals

Animal respiration

Precipitation

Evaporation

Plankton respiration

Plankton photosynthesis

Released by plants and animals into water

Oxidation of organic sediment

fuel burning

Lost from oceans

Dissolved in oceans

Added to oceans by aquatic plants

Limestone forms

Fossil fuel extraction

The oxygen content of the Earth's atmosphere is unusually high, 21 percent by volume, compared with other planets. This is because oxygen not only supports life but arises from life.

When photosynthesizing plants evolved in the oceans 2,000 million years ago, atmospheric oxygen levels began to increase and reached present levels 400 million years ago. Since then the oxygen cycle has maintained a balance between production by plant photosynthesis and consumption by animal respiration.

Humans are interfering with the oxygen cycle by releasing chemicals which upset ozone-oxygen interactions in the stratosphere. Global oxygen levels are under threat from fossil fuel burning, in which carbon is converted to carbon dioxide; from forest clearance, which means photosynthesis is reduced; and from pollution damage to oceanic phytoplankton.

The never-ending story

Our planet is teeming with plant and animal species, but we know virtually nothing about most of them. Nearly 2 million species have been identified so far but there may be as many as 20 million species on Earth. The number and diversity of species increase latitudinally from the relatively barren poles to the bulging equator. Tropical rainforests form the richest ecosystem on Earth and contain 40 percent of all the different animals and plants on the planet, including 4,000 species of tree alone.

The tropics receive greater amounts of sunlight, which allows rapid growth of plants – the basis of all food chains. So the greater the plant growth, the greater the diversity of animal species. In addition, middle and high latitudes have experienced frequent periods of intense disturbance of climate, habitats and soils when thick ice sheets covered large areas of the northern continents. By contrast, the tropics were largely unaffected during these ice ages.

A collection of animals, plants and microbes that live and interact together, sharing the

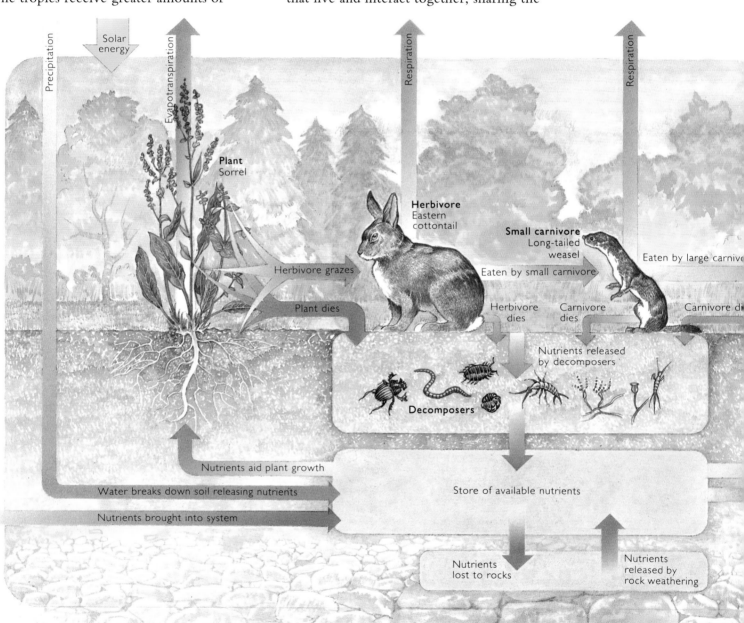

Precipitation

Solar energy

Evapotranspiration

Respiration

Respiration

Plant Sorrel

Herbivore Eastern cottontail

Small carnivore Long-tailed weasel

Eaten by large carnivore

Herbivore grazes

Eaten by small carnivore

Plant dies

Herbivore dies

Carnivore dies

Carnivore d

Nutrients released by decomposers

Decomposers

Nutrients aid plant growth

Water breaks down soil releasing nutrients

Store of available nutrients

Nutrients brought into system

Nutrients lost to rocks

Nutrients released by rock weathering

same environment, is called a community. It is held together by flows of energy and nutrients from the eaten to the eater. Together with its physical surroundings, each community comprises an ecosystem. This can range from a pond or tree to a forest or ocean.

Ecosystems are self-regulating. If favourable weather allows vegetation to grow in abundance, the population of a grazing species such as zebras will increase. These greater numbers will in turn support more carnivores to hunt them, checking the zebra population.

While the zebras are numerous they eat more grass, reducing their food supply and thereby limiting the number of zebras that can be supported.

In other words, a change at one point in the ecosystem causes a response at another point which limits the change and restores the situation to its former state. Restoration can sometimes take several years, often with a marked see-saw of population explosions and declines as happens when desert locusts swarm or large numbers of lemmings go on the march.

Respiration

Animals move in and out of ecosystem

Large carnivore Bobcat

Nutrients taken out of system

Nutrients leached out of system by groundwater

Sun and precipitation enter an ecosystem and trigger flows of energy and nutrients. Plants trap sunlight with the help of chlorophyll. They break water into hydrogen and oxygen, release some of the oxygen, and combine the hydrogen with carbon dioxide to make sugars, in a process known as photosynthesis. Proteins are made from the molecules of chemical salts absorbed in solution through their roots.

Plants are the self-nourishing primary producers of an ecosystem. Herbivores obtain the chemical energy they need by eating the plants, flesh-eating carnivores by eating the herbivores. Each trophic level – plants, herbivores, carnivores – transfers only 10–20 percent of the energy it fixes to the next level. This large loss of energy with each transfer means that relatively few carnivores can be supported.

Unlike energy, which comes from sunlight and leaves as heat, nutrients are regenerated, recycled and retained within the ecosystem. Plant and animal remains are decomposed by bacteria, microorganisms and fungi and converted to nutrients in the soil, and the cycle begins again.

In a pyramid of glass and steel covering 2¹/₂ acres (1 ha) of desert in Arizona is a miniaturized Earth with desert, marsh, rainforest, savanna and an ocean 26 feet (8 m) deep.

This airtight structure is Biosphere 2, a working model of our self-sustaining planetary ecosystem. It supports 3,800 plant and animal species and eight people.

Soon after the two-year experiment began in 1991 carbon dioxide levels inside rose, reaching eight times those outside, and a mechanical means had to be used to reduce the gas.

This study will help plan for space colonies on the Moon or Mars.

The never-ending story/2

An ecosystem changes over time, moving through a series of stages in a process known as succession. A community that develops on land recently exposed by the sea, uncovered by ice or formed from an infilled lake slowly alters its habitat, accumulating more energy and nutrients within it. This eventually leads to its replacement by a different community – one which can make fuller use of the increased resources. Ultimately, a climax or self-sustaining mature stage is reached, such as a tropical rainforest or temperate deciduous forest, in which there is a wide diversity of species and complex food chains.

Species do not stay constant; competition in an ecosystem can make them change. Some individuals display minor differences in their behaviour from the "norm" of their species, especially in connection with obtaining food. Over time, if these individuals thrive and breed better than others, a new species can evolve as their advantageous differences are passed from generation to generation.

Competition can create species with small but highly significant differences in their lifestyle. Three species of brightly coloured songbirds called tanagers coexist on the island of Trinidad even though they eat the same insects from the same trees. The speckled tanager eats insects only from leaves, the bay-headed tanager from the underside of large branches, and the turquoise tanager only from fine or dead twigs.

Natural ecosystems are remarkably resilient but human interference has become so widespread that many can no longer adapt quickly enough. Increased farming, forestry, mining and urbanization together with pollution of the air, rivers, oceans and land are damaging the habitats of many ecosystems beyond repair, even causing the extinction of some species.

If destruction of the tropical rainforests continues at the current rate, several hundred species of vertebrates and over a million species of insects will be extinct by 2030. Greater consideration needs to be given not only to how such damage will adversely affect the planet in the future but also to what can be done to prevent this damage in the first place.

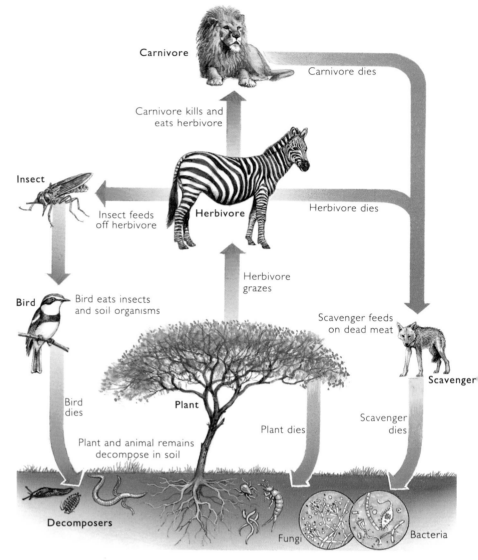

In a complex food web zebras are a part of the herbivore community. They graze in vast herds on the grassy plains of Africa and provide food for carnivores such as lions, cheetahs, hyenas and leopards.

Jackals, vultures and marabou storks scavenge the leftovers from carcasses. When the animals die their remains decompose as a result of the actions of soil bacteria and microorganisms and the nutrients are recycled.

The food web also includes insects such as horseflies and tsetse flies, which feed off herbivores. These insects provide sustenance for a variety of different birds, including the Senegal puffback flycatcher.

Some birds may also feed off the creatures found in topsoil, which are responsible for breaking down decomposing material. Trees, for example acacias, and other plants make use of these nutrients to grow, and in turn are eaten by herbivores, beginning the cycle again.

When the food web of a balanced ecosystem is disturbed problems may arise, with ramifications at all levels. Food supply shortages can be triggered by drought, disease, a population explosion in one species or poaching.

A jackal guards the remains of a baby zebra as vultures close in for their share of the fresh kill. Jackals are scavengers that normally wait patiently and make do with the leftovers of lion, tiger and leopard kills. But some use their superb vision and well-developed sense of smell to seek out calving antelopes in large herds, or even farm herds of cattle, sheep and goats. With its fast reactions the jackal rushes in and snatches a newborn.

Some jackals, desperate for food, can "play dead", attracting birds or small carrion feeders which are then attacked with lightning speed.

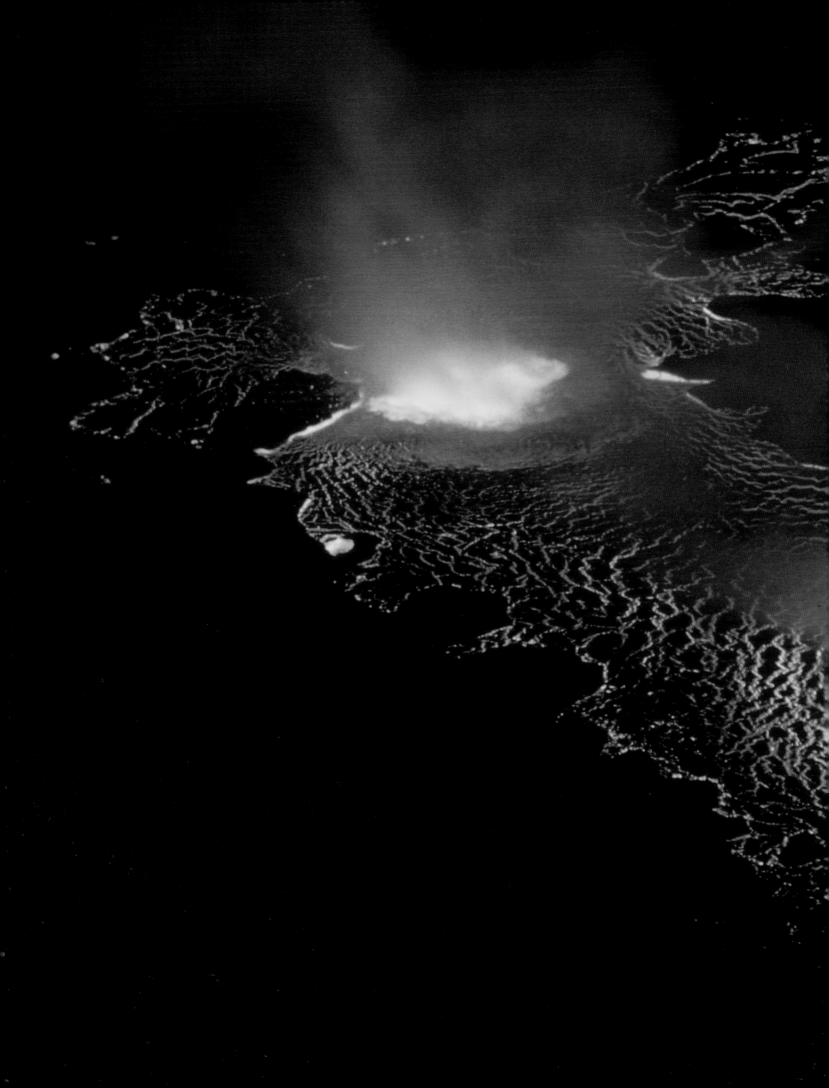

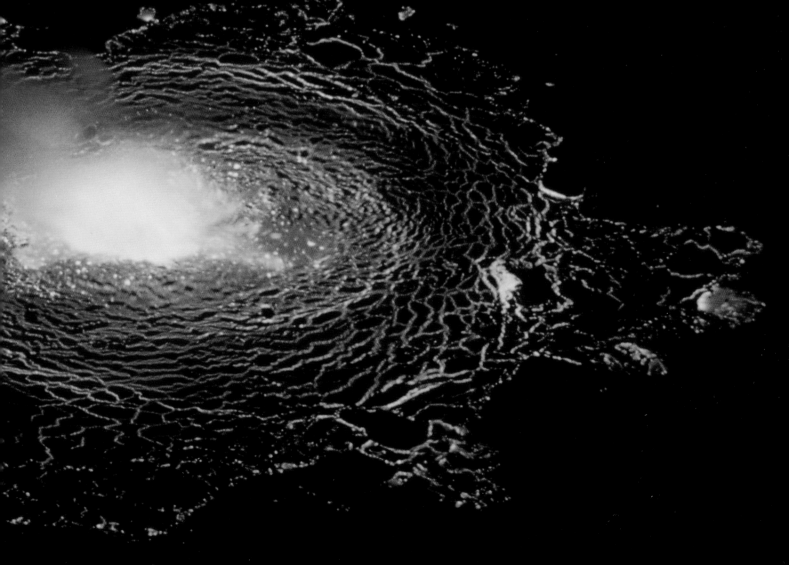

THE INTERIOR

The interior

Energy from the blazing Sun is the key to life in the oceans and atmosphere but it is heat from within the Earth that shapes the continents and ocean basins. Slowly rising and subsiding heat currents inside the Earth control the way in which the giant jigsaw pieces of the planet's crust fit together.

Plumes of geothermal heat orchestrate a slow-motion spectacular – at the speed at which a fingernail grows – in which huge sections of the Earth's crust are torn apart, driven together or scraped against each other. The planet's surface changes radically over hundreds of millions of years as plate collisions thrust up mountain ranges and rip open deep ocean trenches.

Some 200 million years ago the world was a very different place: the Atlantic Ocean did not exist and all the continents were linked together. By the time Africa became the cradle of human evolution only two million years ago, Australasia and the Americas were remote and isolated. As a result they have only been populated by humans for the last 30,000 years. Had human evolution occurred 200 million years ago both these continents, being joined to the others, would probably have played a more significant role in human development.

The cause of the widening oceans and wandering continents is found beneath the Earth's crust. Vast amounts of energy are generated in the heart of the planet by a roaring nuclear-powered furnace. Variations in heat and pressure create a series of layers within the Earth. At its centre is an intensely hot nickel–iron solid core which merges into a viscous liquid core and, eventually, gives

Gases bubble up through whirlpools of superheated lava which rise from the depths of Nyiragongo, in Zaire's Virunga mountains (previous page). This volcano, which stands 11,385 feet (3,470 m) high, last erupted in 1982.

way to an outer layer of semimolten magma, the mantle. This is topped by the surface crust, which is solid but extremely thin: if the Earth were the size of an apple, the skin of the fruit would represent the thickness of the crust. The oceanic floor crust is typically only about 4 miles (6 km) thick, while the less dense continental crust is slightly deeper, averaging 20 miles (32 km).

Magma welling up within the mantle may burst through the surface as lava in spectacularly explosive volcanic eruptions or massive viscous outpourings. On its passage to the surface magma can play a profound part in shaping the interior. The tremendous heat and pressure of the magma may transform other rocks into metamorphic rocks or, if it fails to reach the surface, it may solidify into igneous rock.

In yet another of nature's cycles both these rock types can be weathered into fine sediments which in turn are washed into the oceans, compressed and cemented by percolating fluids into layered sedimentary rocks. Such rocks often hold valuable clues to the nature of life millions of years ago. Remains of plants and animals living at the time the rocks were laid down may be sandwiched in the layers of rock, perfectly preserved as fossils, or they may have decomposed leaving behind coal, oil and natural gas.

Sedimentary rocks laid down on the ocean floor can later be uplifted and contorted into mountains; they can also be heated and melted into metamorphic rock or even be absorbed into the mantle to begin the rock cycle again. Rocks can be transformed and their minerals recycled many times which explains why there are so few rocks recognizable as dating from Earth's early history, say 3,500 million years ago.

Although it covers the entire planet, the Earth's crust is neither seamless nor stationary, but is fragmented into mobile semi-rigid plates. The seven giant plates and many smaller ones jostle one another continuously as they are constantly moved around by the powerful, restless convection currents of the underlying mantle.

New ocean crust is created either side of an ocean ridge where the plates are being driven apart by an upwelling. As the sea floor tears apart, magma rises to plug the gap. It cools and solidifies into large underwater mountain ranges each side of the ridge, such as along the Mid-Atlantic Ridge. Old crust is destroyed where a plate is forced beneath, or subducted by, a more powerful plate, melting as it sinks into the underlying magma. This process can create deep ocean trenches as well as long series of mountain chains.

Plate boundaries are violent zones, with the margins prone to fracture, deformation and melting. When the build-up of molten magma is particularly great explosive plumes of molten lava and jets of gases break through to the surface. Eruptions vary in intensity: lava may shoot out in solid blocks or be sprayed up in a fountain or gush forth in massive fluid flows. Alternatively, they may be cataclysmic explosions emitting ash, lava blocks and glowing avalanches. Whatever their nature, these violent events are not restricted to plate margins.

Large plates can float over localized hot spots — narrow columns of faster-rising magma — and a succession of volcanic eruptions can burst through the middle of a plate. A hot spot, which may be fuelled for tens of millions of years, can produce a line of volcanoes, for example along the Hawaiian Islands-Emperor Seamounts chain in the Pacific Plate. The farther away from the hot spot the older the volcano.

Plates sliding past one another create fault lines, such as the notorious San Andreas Fault in California. The scraping of plates causes strains and stresses which continue to build up until a sudden release occurs, shifting the crust on either side of a fault and triggering an earthquake, often with devastating results. An awareness of the nature of plate boundaries in such populated areas can become a matter of life and death.

An improved understanding of how, where and when crustal plate motion triggers earthquakes and volcanoes, with their associated hazards, including landslides, lava flows, mudflows and seismic sea waves, or tsunamis, will enable us eventually to predict them. There is much to be gained from increased knowledge of the processes operating within the Earth's interior, and their links with the surface, since they may not only help us shape the future but also interpret the past.

The inside story

When Jules Verne wrote *Journey to the Centre of the Earth* in 1864, there were many conflicting theories about the nature of the Earth's interior. Some geologists thought that it contained a highly compressed ball of incandescent gas, while others suspected that it consisted of separate shells, each made of a different material.

Today, well over a century later, there is still little direct evidence of what lies beneath our feet. Most of our knowledge of the Earth's interior comes not from mines or boreholes, but from the study of seismic waves – powerful pulses of energy released by earthquakes.

The way that seismic waves travel shows that our planet's interior is far from uniform. The continents and the seabed are formed by the crust – a thin sphere of relatively light, solid rock. Beneath the crust lies the mantle, a very different layer that extends approximately halfway to the Earth's centre. Here, the rock is the subject of a battle between increasing heat and growing pressure.

In its higher reaches, the mantle is relatively cool and, together with the crust, it forms a solid region known as the lithosphere. At greater depths, in the asthenosphere, or "weak sphere", high temperatures make the rock behave more like a liquid than a solid. Deeper still, in the mesosphere, the pressure is even more intense. This pressure prevents the rock from melting in spite of a higher temperature.

Beyond about 1,800 miles (2,900 km), a great change takes place and the mantle gives way to the core. Some seismic waves cannot pass through the core and others are bent by it. From this and other evidence, geologists conclude that the outer core is probably liquid, with a solid centre. It is almost certainly made of iron, mixed with smaller amounts of other elements such as nickel.

The conditions in the Earth's core make it a far more alien world than the space that surrounds us. Its solid iron heart is subjected to a pressure of about 3 or 4 million atmospheres, and probably has a temperature of about 9,000°F (5,000°C). Although we can speculate about its nature, neither humans nor machines will ever be able to visit it.

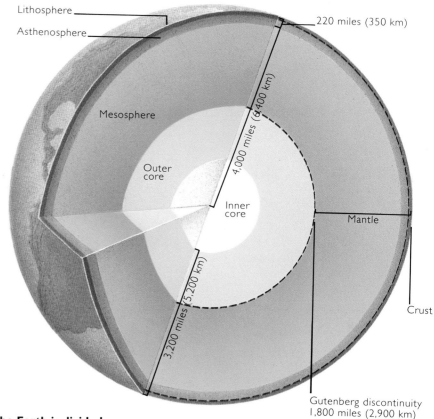

The Earth is divided into different layers like a giant onion, each with its own particular characteristics. The principle that governs this arrangement is a simple one: flotation. If oil, water and syrup are poured into a glass, they will arrange themselves in three layers, with the lightest liquid – oil – floating on the top.

In the same way, Earth's different component materials float one above the other, even though some of the components are solid. The lightest material, continental crust, is known to have a density of about 1.6 oz per cubic inch (2.7 g/cm³).

Calculations of the Earth's mass, first made in 1798, show that the material in the core must have a density of about four times this figure. Iron is the only substance that fits this description.

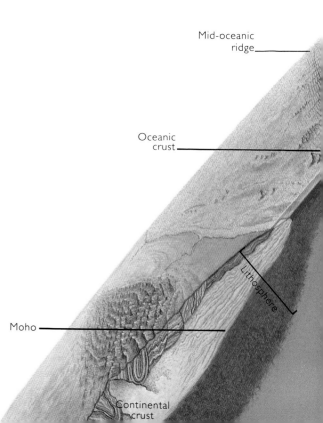

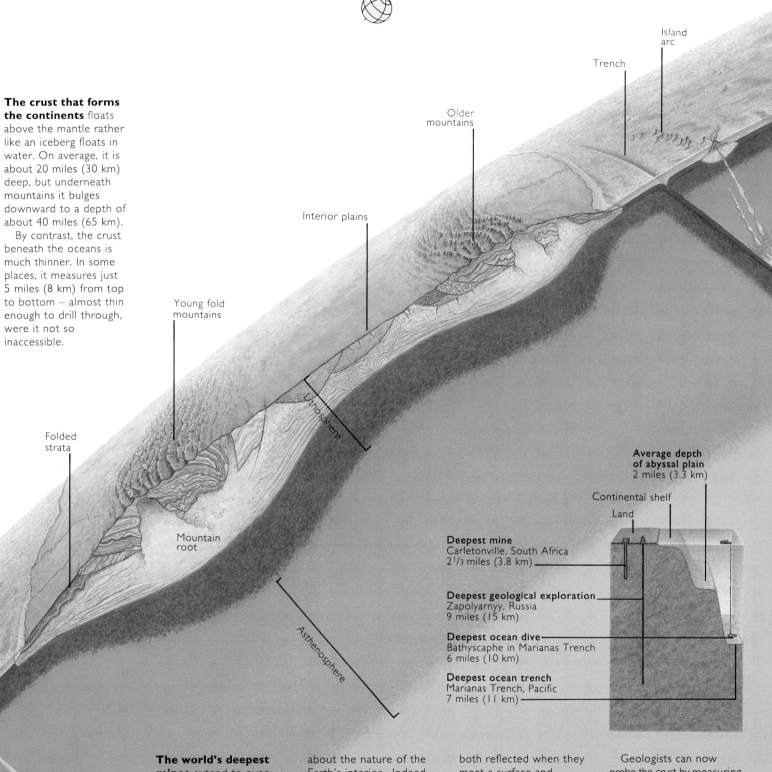

The crust that forms the continents floats above the mantle rather like an iceberg floats in water. On average, it is about 20 miles (30 km) deep, but underneath mountains it bulges downward to a depth of about 40 miles (65 km).

By contrast, the crust beneath the oceans is much thinner. In some places, it measures just 5 miles (8 km) from top to bottom – almost thin enough to drill through, were it not so inaccessible.

Island arc

Trench

Older mountains

Interior plains

Young fold mountains

Lithosphere

Folded strata

Mountain root

Asthenosphere

Average depth of abyssal plain 2 miles (3.3 km)

Continental shelf

Land

Deepest mine
Carletonville, South Africa
2 1/3 miles (3.8 km)

Deepest geological exploration
Zapolyarnyy, Russia
9 miles (15 km)

Deepest ocean dive
Bathyscaphe in Marianas Trench
6 miles (10 km)

Deepest ocean trench
Marianas Trench, Pacific
7 miles (11 km)

The world's deepest mines extend to over 2 miles (3.3 km) beneath the surface, and the deepest boreholes have reached depths of about 9 miles (15 km). They show that average temperatures increase with depth – in deep gold mines, for example, the rock temperature can reach 120°F (49°C) – but they give few other clues

about the nature of the Earth's interior. Indeed, when compared to the Earth's average radius of 4,000 miles (6,400 km), they are mere scratches on the surface.

In 1909, the Yugoslav geologist Andrija Mohorovicic realized that seismic waves could be used to probe the Earth beneath us. Like light waves, seismic waves are

both reflected when they meet a surface and refracted, or bent, when they travel from one medium to another. By comparing the arrival times of sets of seismic waves, Mohorovicic was able to show that there is a sharp boundary between the crust and the mantle, which is now known as the Mohorovicic discontinuity, or moho.

Geologists can now probe the crust by measuring gravity anomalies – small variations in the force that gravity exerts at the surface. Gravity is dictated by mass, so the less dense a rock is, the smaller its gravitational effect. Where the crust is deep, its low-density rock creates a negative gravity anomaly that can be detected from the surface.

The continental jigsaw

During the 16th century, Portuguese cartographers produced the earliest detailed maps of Brazil and revealed an extraordinary geographical coincidence. If the east coast of South America and the west coast of Africa could be brought side by side, the two would fit together like pieces in a jigsaw.

The English philosopher Roger Bacon, writing in 1620, believed that such a strong similarity could hardly be an accident. Others suggested that the biblical flood might have been responsible, its water splitting a giant landmass, pushing the two continents apart.

During the 19th century, as mapping techniques improved, the link between pieces in the "jigsaw" became even more startling. Identical rock formations were found on corresponding coasts thousands of miles apart, and signs of glaciation were discovered on rocks baking under the tropical sun. Matching fossils were found in places as far apart as India and Antarctica, and the plant and animal life of places separated by vast oceans sometimes proved to be extraordinarily similar.

In 1912, Alfred Wegener, a German meteorologist, suggested that the continents, giant islands of light, granite-based rock, somehow ploughed their way through the heavier rock of the sea floor. By running this process in reverse, he concluded that the continents were once joined together in a single supercontinent, which he called Pangaea, that was surrounded by a global sea.

The theory of continental drift was greeted with scepticism. The idea that the continents had moved seemed plausible, but few people could see how solid rock could drift through solid rock. Wegener died in 1930 and his ideas made little further progress.

Then, in the 1950s and 1960s, it became apparent that the Earth's surface was not a single unit, but that it was divided into jagged slabs, now called tectonic or lithospheric plates. Furthermore, these plates were not fixed. Instead, they were moving slowly and inexorably relative to their neighbours. In the space of a few years, plate tectonics revolutionized geology, and continental drift changed from offbeat theory to accepted fact.

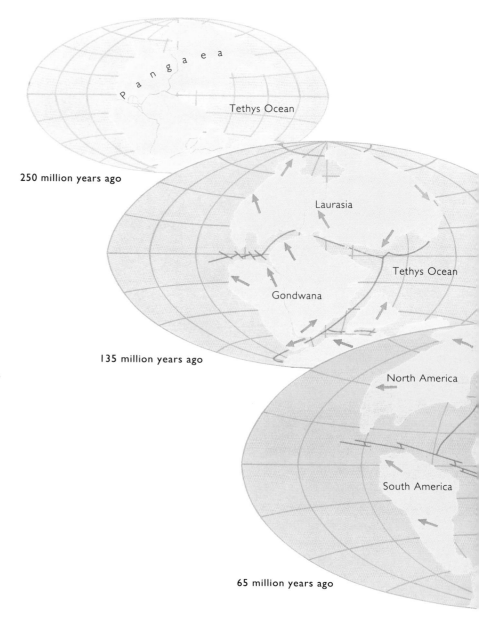

250 million years ago

135 million years ago

65 million years ago

The Earth's continents slide over its surface on board tectonic plates, like participants in an infinitely slow dance. These four maps trace the different steps in that dance, starting 250 million years ago with the formation of the supercontinent known as Pangaea (meaning "all land").

Pangaea formed from three smaller continents, but their amalgamation was to prove only temporary. By 180 million years ago, Pangaea started to break up, as its parts were borne away by plates that were moving in different directions.

Initially, two landmasses were formed – the northerly continent of Laurasia and its southerly counterpart Gondwana. By 65 million years ago, Laurasia and Gondwana themselves had started to break up and the familiar outlines of today's continents came into being. Laurasia split apart to form North America and Eurasia, while Gondwana created India, as well as the continents that currently lie on, or south of, the equator.

Continental drift is a continuous process. It is therefore quite possible that, at some time in the future, the continents will come together once again, and a second Pangaea will be created.

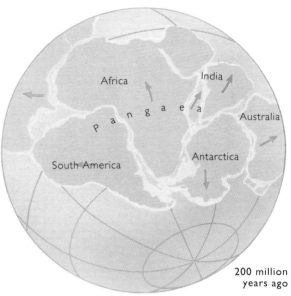

200 million years ago

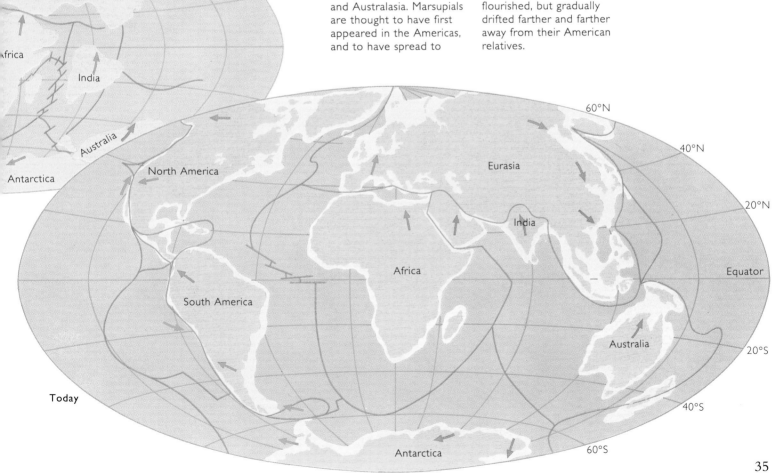

Marsupial mammals, such as this Virginia opossum, are a classic example of "discontinuous distribution" caused by continental drift.

Today, marsupials exist only in two areas that are far apart – the Americas and Australasia. Marsupials are thought to have first appeared in the Americas, and to have spread to Australia via Antarctica, at a time when all these continents were joined.

When the continents broke apart, Antarctica was carried southward, and its marsupials gradually perished as it grew colder. The Australian marsupials flourished, but gradually drifted farther and farther away from their American relatives.

The position of the continents 200 million years ago. A large wedge of continental shelf flares eastward from the tip of South America, filling the apparent gap between that continent and southern Africa.

Today

The continental jigsaw/2

If an egg could be broken open and the pieces of shell fitted together again, it would give some idea of how tectonic plates cover the Earth's surface. Like pieces of eggshell, tectonic plates are rigid. Each one is a slab of lithosphere – crust glued to a layer of solid mantle – with edges that run like jagged cracks over the Earth's surface. Unlike eggshell "plates", the shape of tectonic plates can slowly change, and this allows them to move.

The science of plate tectonics is still in its infancy. The first detailed map of the Earth's plates appeared in 1968, and geophysicists are still uncertain about plate boundaries in some parts of the world. However, it is clear that there are seven large plates as well as smaller ones, and that all are moving relative to each other at different speeds.

Tectonic plates do not propel themselves. Instead, they are probably dragged along like giant rafts by convection currents that circulate within the Earth. The "upside" of each current is hot, molten rock, or magma, that rises up from the asthenosphere beneath the sea floor.

The magma squeezes up through the boundary between plates, helping to move them apart and creating new crust on the seabed. As this process continues, successive bands of newly formed crust join a slow procession away from the plate boundary and across the sea floor.

Eventually, after a journey that lasts up to 200 million years, they meet the lithosphere of a neighbouring plate, moving in the opposite direction. Now the "downside" of the current comes into the effect. One of the plates is subducted, or overridden, by the other, and the cold, dense crust, still glued to its solid mantle, sinks downward, tugging at the rest of the plate behind it. As it moves downward it melts, and the process turns full circle.

Continents are like bulky and elderly passengers, carried on the solid mantle. Their rock is too light to be driven down when plates collide, and instead, it remains at the surface, often exchanging one "raft" for another in the process. The persistence of continental rock means that it can be very old indeed, far older than the rock of the sea floor.

Tectonic boundaries skirt the Earth, dividing its surface into a mosaic of rigid plates that move toward, alongside or away from their neighbours. Satellite measurement systems show that the North American and Eurasian plates, which meet at the Mid-Atlantic Ridge, are separating from each other at a rate of just over 1 inch (2.5 cm) a year.

In the Pacific Ocean, the Nazca and Pacific plates, which meet at the Eastern Pacific Rise, are separating at about six times that speed. Since Europeans first arrived in South America, five centuries ago, the two plates have slid away from each other by 250 feet (75 m).

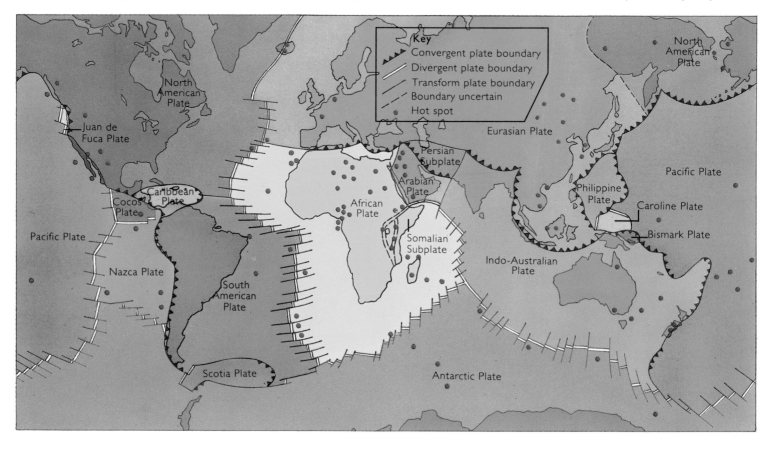

Key
- Convergent plate boundary
- Divergent plate boundary
- Transform plate boundary
- Boundary uncertain
- Hot spot

North American Plate
Juan de Fuca Plate
North American Plate
Eurasian Plate
Pacific Plate
Persian Subplate
Arabian Plate
Philippine Plate
Caroline Plate
Caribbean Plate
Cocos Plate
African Plate
Somalian Subplate
Bismark Plate
Pacific Plate
Indo-Australian Plate
Nazca Plate
South American Plate
Scotia Plate
Antarctic Plate

The Mid-Atlantic Ridge is a divergent or spreading boundary, where new crust is being produced. As the plates slide away from each other, magma wells up between them to form new sea floor. The ridge's central zone is marked by a deep rift valley, similar to those on land.

The Himalayas were formed by a collision between two plates bearing continental crust. When such collisions occur, the continental crust remains at the surface, but is squeezed and buckled by plate movements. Eventually, the movement ceases, and the plate boundary becomes tectonically inactive.

The Japan Trench, off the island of Honshu, marks the point where the Pacific Plate is being subducted by, or driven under, the Eurasian Plate. Here, tectonic activity creates upwelling magma that builds into volcanoes, and also triggers earthquakes.

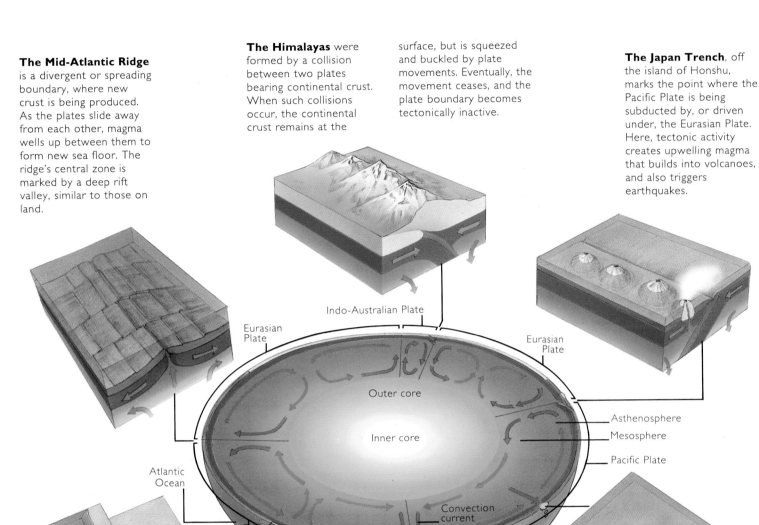

Indo-Australian Plate

Eurasian Plate

Eurasian Plate

Outer core

Inner core

Asthenosphere

Mesosphere

Pacific Plate

Atlantic Ocean

Convection current

North American Plate

Pacific Ocean

Pacific Plate

California's San Andreas Fault is a transform boundary, a region where two plates are sliding past each other, without any crust being created or destroyed. The plate movement often offsets surface features such as fences and roads. Similar faults are found on the ocean floor.

There are a number of theories about the nature of convection currents in the mantle. Some suggest that the currents circulate throughout the whole of the mantle's depth, while others hold that only the asthenosphere is involved.

Oceanic crust
Continental crust
Lithosphere
Asthenosphere
Convection current
Direction of crust movement

The Hawaiian Island chain has been created by a hot spot, a spring of molten magma below the crust. As the Pacific Plate travels northwest, magma bursts through the crust to produce a string of volcanic islands. Away from the hot spot each island's vulcanism dwindles and then ceases.

The continental jigsaw/3

No one knows how long the Earth's continents have been on the move or, indeed, when they first appeared. Ancient, crumpled rocks in northern Canada show that continents were colliding as long as 2,000 million years ago, and geologists believe that at least half of the continental crust was already in place 500 million years before that. Continental drift is therefore a very ancient phenomenon.

For the inhabitants of southern California and Mexico's Baja California peninsula, the relative movement of neighbouring plates is of particular interest. Nearly all of North and Central America is borne by a single continental plate, but these two regions are part of the Pacific Plate, a giant sheet of ocean floor that stretches west as far as Japan. Relative to the North American Plate, the Pacific Plate's margin is moving northwest at an average speed of 2 inches (5 cm) a year.

In a million years, the movement of the Pacific Plate will carry this slice of land 30 miles (50 km) to the northwest. Within 10 million years, San Francisco will be flanked by Los Angeles, travelling up the coast. After a further 50 million years, southern California and Baja California may be torn away from the mainland, as they move toward their final destination, Alaska. Here, the Pacific Plate is being subducted. The drifting land will be scraped from the plate that carries it and become part of the mainland once more.

The supercontinent of Pangaea broke up through a process called rifting, in which slabs of land pulled away from each other. Today, the same process is occurring in Africa. Because the plate boundaries in this region are complex, it is difficult to predict what might happen.

However, if the Horn of Africa continues moving northeast, within 50 million years a new sea will open up in east Africa, allowing salt water to flood across the continent up to Lake Victoria. The Persian Gulf will disappear as the Arabian Plate collides with the Eurasian Plate, and the Red Sea will widen and link up with the Mediterranean. The entire region, which has been so important in the history of humanity, will change beyond recognition.

The San Andreas Fault, which slices through coastal California, marks an uneasy boundary where one tectonic plate is sliding past another. In this view of the desert north of Los Angeles, the fault crosses the Carrizo Plain.

The land in the foreground is perched on the Pacific Plate, while that beyond is shouldered by the North American Plate. As the Pacific Plate slides slowly to the northwest, it will eventually carry part of California out to sea.

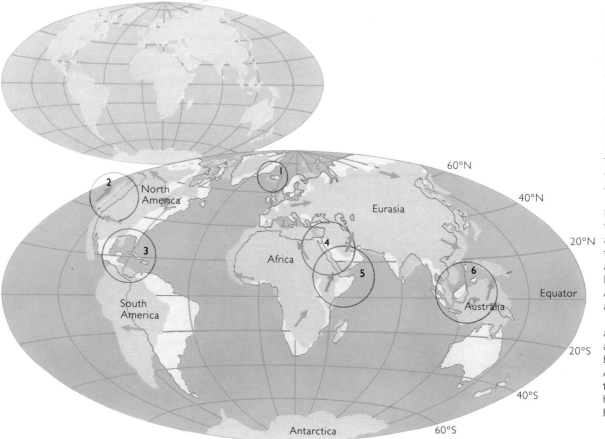

Evidence from the Earth's crust shows that continental drift has always taken place at roughly the same rate. This map shows the projected position of the continents 50 million years from now. The Atlantic Ocean has widened and Iceland, which lies astride the Mid-Atlantic Ridge (1) continues to get wider by about 1 inch (2.5 cm) a year.

On the west coast of North America, a fragment of California is about to tear itself free of the mainland (2). North and South America are no longer joined (3), and Africa and Asia have split apart at the Red Sea (4).

East Africa is splitting along the Great Rift Valley and may eventually become an island (5). Australia has continued to move northward and has collided with Indonesia (6).

Ocean ridges and trenches

Until 150 years ago, very little was known about the nature of oceans. Nautical charts showed depths close to coasts, but few measurements had been made far out to sea. Taking soundings with a weighted line was slow and laborious, and there was little to gain in fathoming the oceanic deeps.

But in the 1850s, a scheme was put forward to lay a submarine cable from Europe to North America. Matthew Maury, an American naval officer, assembled all the known depth data for the North Atlantic, and discovered that the ocean's central section – which he called the "Telegraph Plateau" – was much shallower than the broad regions that flanked it. In the 1870s, oceanographers aboard the British research vessel *Challenger* discovered similar features in the Pacific, and gradually it became clear that the floor of the open oceans rose and fell just as precipitously as dry land.

Modern imaging methods, which include sonar and satellite-borne sensors, show that a system of oceanic ridges runs around the Earth, making up the largest continuous range of mountains on our planet. Some of these ridges have deep valleys at their centres, while others have rocky spines. There are also transform faults – scarlike gashes at right angles to the ridges – and oceanic trenches,

Throughout history, sea levels have risen and fallen, and the edges of continental shelves, rather than today's changing coastlines, mark the true frontiers between the continents and the oceans.

Each ocean is made up of a number of plates, which meet at ridges and trenches. Trenches always lie close to land, but ridges may be central in an ocean, or offset to one side.

The sea floor is constantly regenerating. The cycle begins at an oceanic ridge, where magma rises to the surface. The new crust is relatively light and rides high above the ocean floor, forming mountains running parallel to the ridge.

As the crust slides away, it begins to cool and sink, and the continuous rain of sediment filtering down from above begins to obscure its outlines.

By the time the crust reaches an oceanic trench, to be subducted by a neighbouring tectonic plate, the sediment may be ½ mile (1 km) thick.

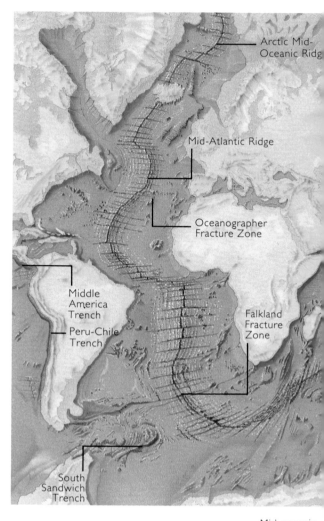

Arctic Mid-Oceanic Ridge

Mid-Atlantic Ridge

Oceanographer Fracture Zone

Middle America Trench

Peru-Chile Trench

Falkland Fracture Zone

South Sandwich Trench

How the sea floor spreads

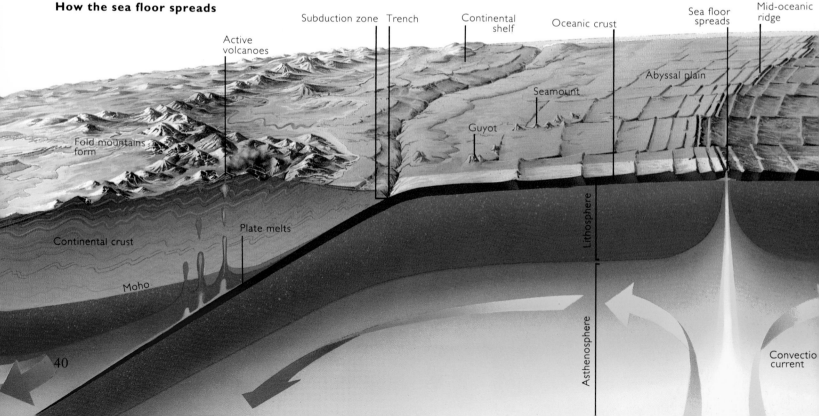

Subduction zone

Trench

Continental shelf

Oceanic crust

Sea floor spreads

Mid-oceanic ridge

Active volcanoes

Abyssal plain

Seamount

Guyot

Fold mountains form

Continental crust

Plate melts

Moho

Lithosphere

Asthenosphere

Convection current

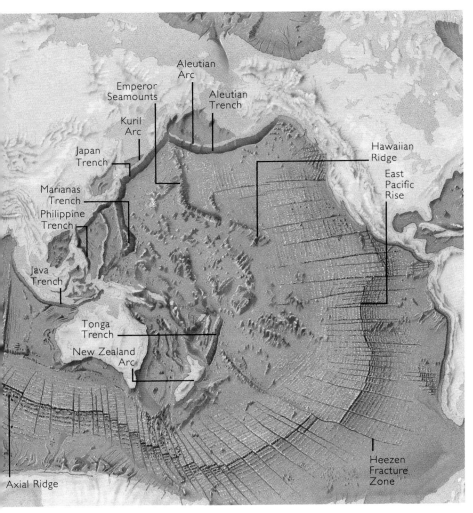

Emperor
Seamounts

Aleutian
Arc

Aleutian
Trench

Kuril
Arc

Japan
Trench

Hawaiian
Ridge

East
Pacific
Rise

Marianas
Trench

Philippine
Trench

Java
Trench

Tonga
Trench

New Zealand
Arc

Heezen
Fracture
Zone

Axial Ridge

where the ocean floor plunges down almost 7 miles (11 km). Between ridges and trenches are abyssal plains – enormous empty expanses cloaked by deep ooze and studded with sea-mounts sometimes reaching as far as the surface.

A ridge marks the site where new crust is squeezing its way to the surface, and a trench the place where it is being engulfed once more. If the two processes of crust creation and destruction are exactly balanced, an ocean maintains its size. But if they are not, the sea floor expands or contracts. Over millions of years, this process brings whole oceans into existence and presses others to extinction.

The Pacific is probably the world's oldest ocean. It dates back about 600 million years, and has an area almost as large as all the other seas and oceans put together. Its floor is exuded by two main ridges, the East Pacific Rise and the Pacific-Antarctic Rise. Once formed, the crust is shunted over the seabed, a few inches every year, until it is eventually consumed by two systems of trenches, one bordering South and Central America, and the other stretching in broken arcs from Tonga, through the Far East, to Alaska. By the time the crust has completed its journey from the ridge to the ocean's western edge, it has reached an age of about 170 million years.

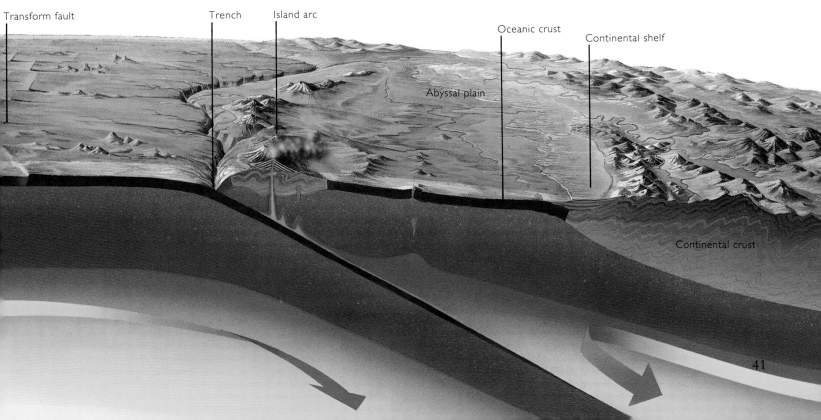

Transform fault

Trench

Island arc

Oceanic crust

Continental shelf

Abyssal plain

Continental crust

Ocean ridges and trenches/2

If the world's sea water could be temporarily removed, few of the scenes visible from the exposed sea floor would rival that of Mauna Kea and Mauna Loa, rising massively out of the Pacific. These twin peaks, which crown the island of Hawaii, reach heights of over 13,000 feet (4,000 m), but their true bulk is normally masked by water. Their mountain base, bigger than Everest, rises gently from the sea floor to a height of 33,000 feet (10,000 m),

sprawling over an area of 250 miles (400 km).

Hawaii was created by magma bursting through the sea floor above a highly active hot spot in the Earth's mantle. Formed less than a million years ago, it is just one of many isolated volcanic mountains that punctuate the world's oceans. Many lesser seamounts tower above the sea floor, but stop within a few hundred feet of the surface.

To date, oceanographers have mapped over

Scoured by the waters of the ocean, Hawaii's jet black sands bear witness to the island's volcanic origins. Hawaii is the most easterly and the most volcanically active island in the Hawaiian-Emperor chain, which stretches more than 3,000 miles (5,000 km) across the northern Pacific.

The moving ocean floor is carrying Hawaii to the northwest, away from the hot spot that created it. Eventually, Hawaii's volcanic activity will cease, and the eroded island will disappear beneath the waves to become a seamount. Meanwhile, the stationary hot spot will form further islands, each destined to share the same fate.

The new island of Surtsey appeared on November 16, 1963, during a spectacular duel between magma and water off Iceland's southern coast.

Surtsey lies over the Mid-Atlantic Ridge, an area where adjoining tectonic plates are slowly moving apart and forming new crust. Unlike Helmsey, its short-lived sister island, Surtsey has proved to be permanent, and is now colonized by plants and seabirds.

2,000 seamounts, mostly in the Pacific Ocean. Studies of these submarine peaks show that they probably represent stages in a special cycle of creation and destruction that takes place both above and below the water's surface.

The cycle begins when a hot spot releases magma, creating a volcano that forms new land. Over millions of years, wind and water erode the island's surface, and at the same time, the crust carries it away from its magma source, into ever deeper water. As the eroded mountain sinks, it may become encrusted with coral, forming an atoll or ring-shaped reef.

But eventually the upward growth of the coral stops and it dies. The result is a guyot, a seamount with a distinctive flat top of coral and sediment. The final stage in the cycle occurs as the seamount is carried into a deep-sea trench. Here it is destroyed, together with the crust that has carried it across the ocean.

Continental stress

When two vehicles collide at high speed, their energy of movement is transformed into forces that can shatter glass and make metal crumple. When continents collide, the same kind of forces are unleashed, but this time on an unimaginably larger scale.

A continental collision lasts not for seconds, but for millions of years. During this lengthy impact, part of the Earth's crust is squeezed more and more tightly until there is only one direction in which it can escape – upward.

When continents smash into each other, the heavy crust that forms the ocean floor is slowly carried downward into the Earth's mantle, melted and destroyed. As the crust is forced downward it generates molten rock, or magma, that may rise to form volcanoes. Eventually, after millions of years, the ocean floor closes over completely leaving a giant bank of sediment. Gradually this lighter material becomes more and more compacted, splits and is thrust upward to form mountains.

The newly formed mountain range marks a suture in the Earth's crust – a line where two continental plates have become welded together. The Himalayas flank a recent suture, while mountain ranges like the Urals and Appalachians mark much older collisions.

The great ranges of Eurasia were all created in the same way as the Himalayas but in western North America it is not the same story. These mountains were probably built not by a clash of equals, but by one in which the contestants were of very different sizes.

The evidence for this lies in the giant geological mosaic that makes up the mountains. Altogether, they are thought to contain about 50 quite distinct islands of rock, called terranes, each of which has a characteristic "fingerprint".

Geologists have concluded that these terranes were originally separate landmasses, formed up to 5,000 miles (8,000 km) away from North America's present position. Each landmass was borne east by the movement of a tectonic plate, and was then scraped from the plate when it met the continental shelf of North America. As each landmass collided with the North American Plate, it was propelled, folded and deformed like a car being scraped by a big truck.

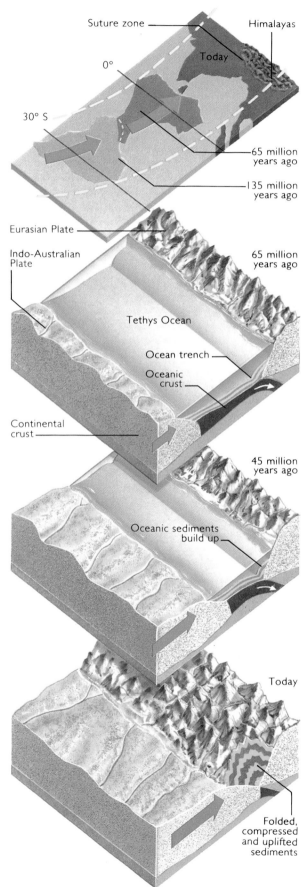

Suture zone · Himalayas · Today · 0° · 30° S · 65 million years ago · 135 million years ago · Eurasian Plate · Indo-Australian Plate · 65 million years ago · Tethys Ocean · Ocean trench · Oceanic crust · Continental crust · 45 million years ago · Oceanic sediments build up · Today · Folded, compressed and uplifted sediments

The most spectacular example of the violence of continental movement can be found high up on the snowy, windswept Tibetan plateau where remnants lie of the great Tethys Ocean which separated two great continents more than one hundred million years ago.

After the global continent of Pangaea broke up some 200 to 250 million years ago the Indo-Australian Plate began to travel northward. It travelled at great speed (in geological terms) and around 45 million years ago smashed into the larger, heavier Eurasian Plate, giving rise to the Himalaya Mountains. The Tethys Ocean floor was propelled forward, up and over the now crumpled margins of the Eurasian Plate, where it sits today.

Geological surveyors have concluded that the Himalayas are a "horrible mess" with strata piled up and squashed together. Deep fault lines have been found where material was first thrust up on collision and then pulled down by gravity.

Even today the momentum is still apparent with the Nanga Parbat Massif of northern Pakistan being pushed upward at a rate of 1/5 inch (5 mm) a year. This means that in a million years, the rise could be as much as 16,000 feet (5,000 m), although erosion will eat away some of this height.

Eventually, the relative movement will cease as the two continents become fused and locked together.

At Stair Hole, on England's southern coast, erosion reveals folds in the underlying rock. These strata were originally laid down horizontally on an ancient seabed. Later deformation has bent and tilted the layers, turning some of them on end so that they run at a steep angle into the sea.

Mount Makalau on the border of Nepal and Tibet is nearly 27,800 feet (8,500 m) high, only marginally lower than the tallest mountain in the world – Mount Everest – which stands to the west of it. On its crest lie the youngest rocks and the ancient seabed of the primeval Tethys Ocean.

45

Continental stress/2

Not all mountains are the result of relentless squeezing as tectonic plates smash together. In the Great Rift Valley, which stretches southward from Turkey to Mozambique, ranks of steep parallel mountainsides have been produced by exactly the opposite effect. Here, slowly but inevitably, the surface of the Earth is being cracked apart.

The Rift Valley marks the point where the African Plate meets the smaller Arabian Plate and the Somalian Subplate. Beneath this long trench, and to either side of it, the continental crust is being heated by molten magma. As the crust expands and swells, neighbouring slabs move away in opposite directions, perhaps aided by tectonic movements. The central span of rock periodically cracks and then collapses, creating precipitous faults, where one block of rock slips downward against another.

The scale of the Rift Valley is staggering. From its beginning in eastern Turkey, it runs due south through the Near East, and then broadens to form the basin of the Red Sea. Here, the submerged valley floor is up to 200 miles (320 km) wide, and its eastern shore is bordered by a plateau with mountain peaks that reach 12,000 feet (3,650 m). The fault block mountains that border the Red Sea's mouth are typical rift valley features; in places they form steps like a giant staircase.

As it reaches Africa, the Rift Valley divides into two lake-filled branches, but the mountain ramparts remain. In places, the faults that border the valley make up cliffs that are hundreds of feet high. The slopes nearest the centre of the valley usually form the steepest section of the escarpment, while those farther away, which mark older faults, are more eroded.

The Rift Valley reaches its end near the port of Beira in Mozambique. Here a question hangs over its future. If tectonic activity has largely ceased in the African section of the valley – as some geologists believe – it will gradually erode. But if the Rift Valley pulls apart once more, this part of Mozambique will be the future scene of a dramatic flood. As neighbouring plates tear away from each other, the coast will be ripped open and the Indian Ocean will pour into the heart of Africa.

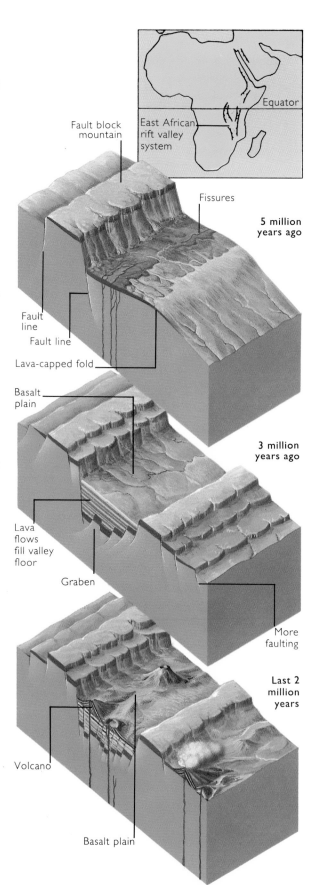

East African rift valley system

Equator

Fault block mountain

Fissures

5 million years ago

Fault line

Fault line

Lava-capped fold

Basalt plain

3 million years ago

Lava flows fill valley floor

Graben

More faulting

Last 2 million years

Volcano

Basalt plain

Hell's Gate in Kenya lies on the floor of the Great Rift Valley, in a strangely enclosed sweep of land bordered by clifflike mountains. Like many parts of the rift, Hell's Gate was once a cauldron of volcanic activity, a sure sign of the thinness of the Earth's crust.

Ancient blocks of lava bake in the tropical Sun, and scattered fragments of obsidian – a black, volcanic glass – can be found on the ground. In the Rift Valley, which slices through Africa's "Cradle of Mankind", early humans used this glass to make tools.

Like folding, faulting is a complicated process, involving many stages. These diagrams show how faulting may have taken place in the African part of the Great Rift Valley.

The valley's life begins when the magma heats the crust, making it expand and bulge outward. Eventually, part of the crust collapses to form a steep wall along a fault line.

Further upward movement of the crust creates more faults, building a series of fault block mountains. In the faulting process the central block of land subsides to form a graben.

The faults in the crust offer an easy way upward for magma. It erupts on to the valley floor pouring out through fissures – to create basalt plains – or through volcanic cones.

Seismic shakes

Of all the natural catastrophes that afflict humankind, earthquakes rank among the most deadly. Yet in some ways, their disastrous effects are partly of our own making. In open ground, where there are no man-made structures, even a severe earthquake can leave little immediate evidence that it has taken place. But if an earthquake of the same power affects a town or city, the scene can be one of terrifying devastation, as buildings topple.

The Earth experiences more than a million quakes a year, most of which are too slight to be noticed. People indoors are usually the first to be aware of an earthquake, because most buildings conduct vibration well, and their contents often break if shaken.

In classical times, the Greek philosopher Aristotle thought that earthquakes were caused by subterranean wind and fire, while in Japan, a country much afflicted by quakes, the shaking ground was interpreted as the thrashing of a giant fish that supported the world. Quakes have also been thought of as a form of divine punishment. When Lisbon

suffered a shattering earthquake on All Saints' Day in 1755, the deaths of over 50,000 people were seen as a rebuke for wrongdoing.

It is now known that the earthquake-prone regions closely follow the zones where neighbouring tectonic plates meet. Minor earthquakes are produced where plates are spreading apart, but much greater quakes occur where plates are sliding past each other, or where one plate is being driven downward, or subducted, by another. During both these kinds of movement, enormous forces are required to overcome the friction between the two plates. In some places, the plates grate past each other with a roughly steady motion, but in others, friction interrupts their progress, and they lock together for months or years.

When this happens, the forces behind the plates build up, distorting the crust much like someone squeezing a spring. The outcome is dramatic. Quite suddenly, something gives way. The two locked plates lurch into new positions, and the energy is released in shock waves that reverberate around the world.

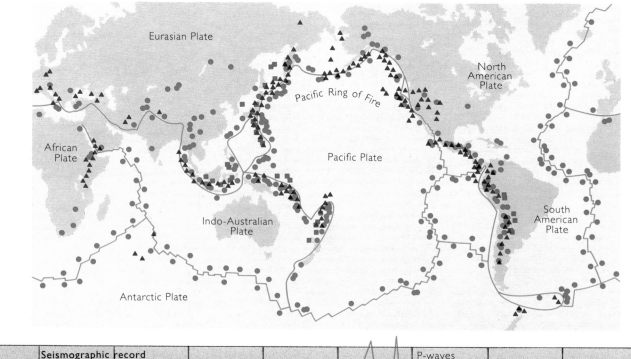

The foci of major earthquakes and the sites of active volcanoes in relation to plate boundaries. An earthquake's focus is the point where its energy is released; its epicentre is the point directly above on the surface.

Most earthquake foci are less than 60 miles (100 km) from the surface, but earthquakes that are triggered by subducted plates can take place at 10 times that depth.

■ Deep earthquakes
● Shallow earthquakes
▲ Volcanoes
╱ Plate margins

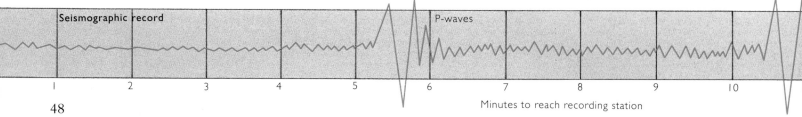

Seismographic record   P-waves   S-waves

1   2   3   4   5   6   7   8   9   10

Minutes to reach recording station

Seismic waves travel at different speeds and along different paths. The first waves to be detected are primary (P) waves. These can travel through the Earth's molten outer core, and they reach the surface on the far side of the globe in about 20 minutes.

Secondary (S) waves are more sluggish, and cannot directly cross the Earth. Instead, they cast a "shadow", revealing where the liquid outer core has blocked their path. As P- and S-waves travel through the Earth, they are gradually bent because the variation in the Earth's density refracts the waves.

Waves can also be reflected when they strike a boundary within the Earth, and a combination of refraction and reflection of P-waves at the boundary between the core and mantle creates a ring-shaped shadow.

An earthquake transmits energy to the surrounding rock in wave movements of four different types. Primary (P) and secondary (S) waves travel through the interior of the Earth, hence their collective name of body waves.

A P-wave moves by compressing and stretching the material that it travels through, while an S-wave makes it shear from side to side. P-waves can travel through solids and liquids, but S-waves can only travel through solids.

Love waves and Rayleigh waves (L-waves), also produced by quakes, are surface effects. They make the ground shear from side to side or oscillate up and down. Rayleigh waves are responsible for the frightening rippling of the ground sometimes seen during a quake.

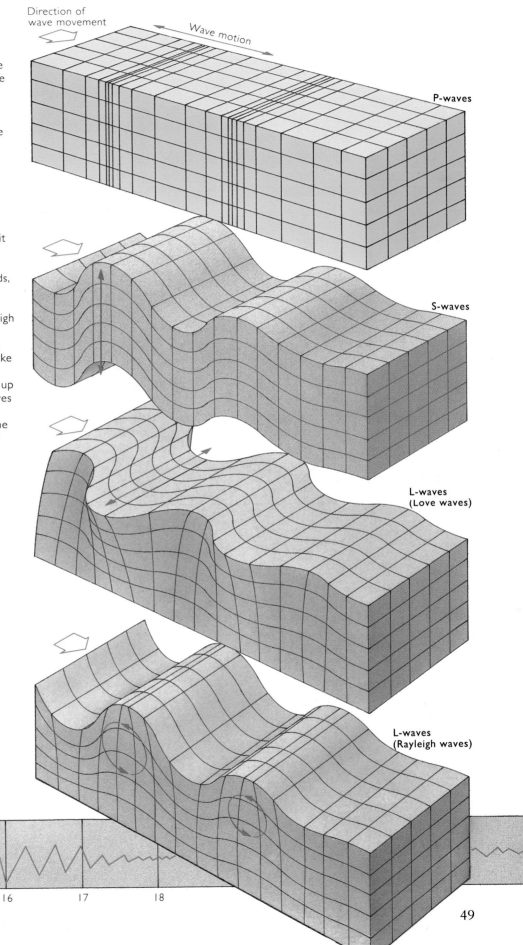

49

Seismic shakes/2

For people who live close to colliding plate boundaries, such as the rim of the Pacific or the Middle East, earthquakes are an unpredictable and unwelcome fact of life.

During this century alone, the Pacific's record has been a grim one. In 1906, an earthquake struck San Francisco, shattering its buildings and bringing about a fire that could not be controlled because water pipes were fractured. More than 700 people died and the city was left in ruins. In the same year, a less publicized quake struck Valparaiso, a coastal city in Chile, killing 1,400 people.

From 1907, severe earthquakes struck Japan for five years in succession, and then in 1923 the city of Tokyo was shaken and set alight with the loss of more than 100,000 inhabitants. On the Pacific's eastern rim, a string of capital cities, including Managua, Mexico City and San Salvador, have all been badly affected.

Mexico City lies over 200 miles (320 km) from the meeting place of the Pacific and North American plates, but the soft lake bed on which it is built reverberated as the earthquake waves arrived in 1985. This undermined the foundations of many structures in the city centre, bringing hotels, hospitals and apartment blocks to the ground, and claiming 10,000 lives.

By contrast, regions far from the active margins of tectonic plates – for example eastern North America, northern Europe or northern Siberia – are much less vulnerable. However, in 1580, an earthquake damaged buildings on either side of the English Channel, and in 1750, a quake caused widespread panic in London, a city where earthquakes are virtually unknown. In 1811 and 1812, a succession of far larger earthquakes shook Missouri and Arkansas, two states that are at the heart of a continental plate, and in 1976, an earthquake levelled the Chinese city of Tangshan, which is situated away from a plate boundary.

These "rogue" earthquakes are, as yet, unexplained. Some may be triggered by ancient plate boundaries hidden deep in continents, while others may be caused by quite different forces, such as the gradual upheaval of the crust following the melting of ancient ice caps.

In this scene of total devastation, a man digs through the remains of an apartment block in Leninakan – now renamed Gumirie – in Armenia. This 1988 earthquake took place near a complex system of plate boundaries where the Arabian Plate is pushing northward into Eurasia.

The quake caused great loss of life because it affected centres of population. Local buildings were unable to withstand the severe vibration, and they collapsed on their occupants.

The San Francisco earthquake of 1989 was a reminder of the dangers of life near a tectonic boundary, and also demonstrated the value of earthquake-proof architecture. Some older buildings suffered severe damage, but the specially designed skyscrapers survived intact.

For many years, Californians have waited for the "Big One" – an earthquake that will match that of 1906. Seismologists monitoring plate movement of the San Andreas Fault suspect that it will be preceded by a seismic silence, when the neighbouring plates lock together. If seismologists can decide exactly how long that silence should last before the earthquake is triggered, earthquake prediction will be a reality.

So far, the investigation of such seismic gaps has produced some accurate predictions, but in general, the results are too imprecise to be useful.

Volcanic fires

On April 2, 1991, the Philippine island of Luzon received its first warning of an impending disaster. Mount Pinatubo, an ancient volcano in the western Zamballes range, stirred from a slumber that had lasted over 500 years. As superheated steam exploded to the surface, ash was thrown high into the air, and the first of hundreds of tremors began to shake the island's western coast.

Pinatubo's reawakening came as a surprise. Filipinos are no strangers to eruptions – in 1976 Mount Mayon, on Luzon's east coast, erupted with the loss of over 1,500 lives – but Pinatubo was no longer thought to pose a threat. Vulcanologists were monitoring many active volcanoes throughout the Philippine archipelago, and there seemed little point studying one destined for extinction.

The April eruption led to a rapid change of priorities. Teams of scientists used laser surveying equipment to record minute changes in the mountain's surface. It became apparent that the initial eruption was just an overture. As a result, a mass evacuation was ordered, and over a quarter of a million people were soon on the move. Their flight was not in vain.

On the night of June 14, Pinatubo ripped itself apart in the greatest volcanic eruption this century. The ash had sufficient energy to punch its way up far into the stratosphere – three times higher than the flight-path of a jet airliner – and was carried westward by the prevailing wind to mainland Southeast Asia. On Luzon itself, much of the western coast was devastated. Ash fell like a searing grey snowfall, killing plants and animals on contact. However, thanks to the prompt evacuation, fewer than 350 people were killed.

Mount Pinatubo's 1991 outburst was large, but it was by no means in the front rank. The explosion in 1883 of Krakatoa, a volcano in the strait between Java and Sumatra, was vastly more destructive. It ejected five times as much material, and killed nearly 40,000 people. The sound waves produced by the explosion could be heard 3,000 miles (4,800 km) away, while the ash circled the Earth.

But even this cataclysm was dwarfed by one that preceded it. In 1815, the eruption of

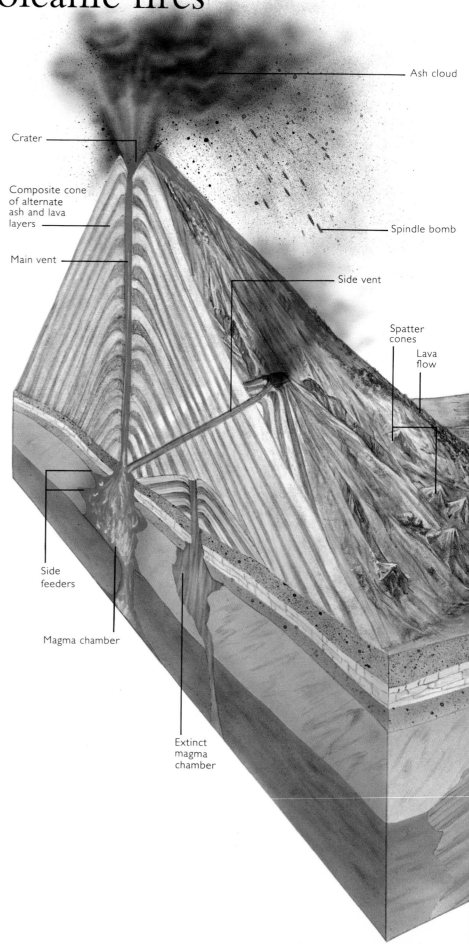

Ash cloud

Crater

Composite cone of alternate ash and lava layers

Main vent

Spindle bomb

Side vent

Spatter cones

Lava flow

Side feeders

Magma chamber

Extinct magma chamber

Volcanoes are created by magma – hot, molten rock that wells up from the mantle to the Earth's surface. This upwelling nearly always occurs at the boundaries between plates or at hot spots.

Far beneath a volcano's vent lies a magma chamber, connected to the upper mantle. Here, magma is confined by the pressure of the solid rock above it. At this depth, it is "squeezed" so intensely that it can hold a large amount of dissolved gas – chiefly carbon dioxide.

But when magma rises through a volcanic vent, the pressure on it falls. The magma can no longer hold so much gas, and it bubbles out of solution. If the rapidly expanding froth of molten rock reaches the surface, the result is an eruption.

The shape of a volcano is determined by the kind of magma it produces. The volcanoes of the Pacific "Ring of Fire" generate andesite magma. Formed where one tectonic plate is subducted by another, andesite is rich in silica. Dissolved gases cannot easily bubble away when andesite approaches the surface. Instead, pressure builds up until the magma blasts its way into the open, immediately cooling to form a shower of solid particles or ash. The ash builds up to produce a conical stratovolcano.

Shield volcanoes are formed from less explosive basaltic magma. This contains much less silica, and it flows away swiftly to build up large, spreading volcanoes with gentle slopes.

Circular craters, at times filled with water, mark the places where explosive eruptions have taken place. Magma leaking from fissures forms flat lava plains, while magma that fails to reach the surface solidifies to form intrusive rock formations, such as dykes and laccoliths.

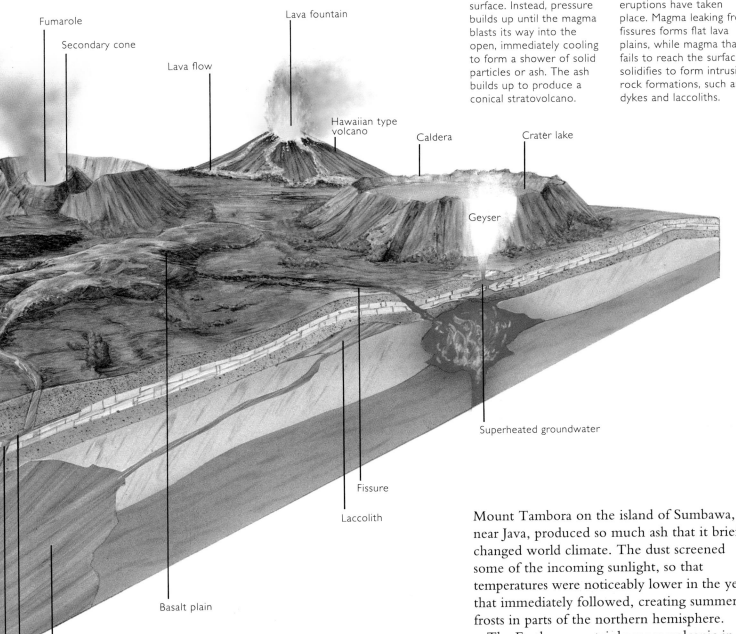

Fumarole

Secondary cone

Lava flow

Lava fountain

Hawaiian type volcano

Caldera

Crater lake

Geyser

Superheated groundwater

Fissure

Laccolith

Basalt plain

Sill

Batholith

Dyke

Mount Tambora on the island of Sumbawa, near Java, produced so much ash that it briefly changed world climate. The dust screened some of the incoming sunlight, so that temperatures were noticeably lower in the years that immediately followed, creating summer frosts in parts of the northern hemisphere.

The Earth was certainly more volcanic in its early history than it is today, but there is no guarantee that truly cataclysmic eruptions are a thing of the past.

Volcanic fires/2

From a distance, a volcanic eruption looks like a giant bonfire pouring smoke into the sky. So it is not suprising that, until the 19th century, many geologists believed that eruptions were indeed fires. But for a fire to burn, there has to be a fuel, as well as oxygen. Despite careful examination of volcanoes by early geologists, no trace of a fuel was ever found.

The fact that volcanoes can erupt out of the sea shows that burning plays little part in the way they work. Volcanoes do produce hot, inflammable gases which catch fire on contact with the atmosphere, but they play only a relatively small part in the inferno of an eruption. The power of a volcano comes from the sudden movement of heat, rather than from its creation.

From a human perspective, the greatest danger that eruptions present is not from the flow of lava, which usually moves more slowly than walking speed, but from material that is ejected into the air. This includes scalding steam, suffocating carbon dioxide, which rolls downhill under its own weight, and a cocktail of rocky particles of many different sizes.

As the particles are flung into the atmosphere, they cool down and solidify. Boulder-sized globules of magma are often shaped by their brief journey through the air, and they plunge to the ground as streamlined and deadly spindle bombs. Smaller particles may form chunks of pumice, a type of rock that is so riddled with cavities that it is light enough to float.

Finally, the smallest particles of all form a fine mineral dust. Collectively known as volcanic ash, these particles usually settle out according to their size, with the lightest travelling farthest. However, sometimes the volcano's steam condenses to form torrential rain, and this washes all the ash to the ground. Here it makes a mudflow, a thick, deadly slurry that envelops anything in its path, and then traps it like fast-setting concrete.

When Mount St. Helens (right) erupted on the morning of May 18, 1980, its summit collapsed as an initial sideways blast flattened trees and triggered a giant landslide that clogged rivers. Minutes later, a column of gas and ash burst upward to a height of about 12 miles (19 km).

The ash was carried east in an ever broadening belt that stretched from Canada to the Caribbean. Within 72 hours this ash cloud had crossed the continent, and in the space of two weeks it had encircled the Earth.

During the eruption Mount St. Helens ejected 1 cubic mile (4 cu km) of airborne debris, and lost 1,300 feet (400 m) of its original height.

Mount Pacayo (left) is one of a number of active volcanoes that lie south of Guatemala City. It is an example of a shield volcano with gently flowing rivers of magma that dribble down to the flat lands below, leaving vast areas of cooling lava that will eventually form a basalt plain.

Fountains of steam

Some 50 miles (80 km) northeast of Reykjavik in Iceland is a wide, treeless valley that leads up to Langjökull, one of the island's ice caps. On the valley's western slopes, by the village of Haukadalur, is the site that brought the word "geyser" into worldwide use. Here, the original *Geysir*, a fountain of superheated water and steam, once erupted high into the air with almost clockwork regularity.

Geysers are powered by geothermal energy – the heat energy that is stored beneath us in solid and molten rock. In most places, the amount of heat that leaks upward is small, because the surface is well insulated by a thick layer of continental crust. However, in Iceland, New Zealand and other places where magma comes close to the surface, the effects of this geothermal heat are very much in evidence.

Because of the extremely high temperatures, no one has been able to explore a geyser to determine exactly how it works, but the principles behind these sudden surges are not too difficult to establish. All geysers are located in areas where the underlying rock is heated by magma. Each geyser probably features an underground chamber, which slowly fills with groundwater seeping in from the surrounding rock. The water begins to heat up, but the confining effect of the chamber, together with its depth, means that the water becomes superheated. At a critical temperature, steam suddenly forms, and the mixture of water and steam explodes upward.

Geysers are often clustered together in dozens or hundreds, with the largest number in Iceland. Wherever they are found so, too, are other geothermal phenomena, including hot springs, pools of boiling mud, and vents called fumaroles.

In geological terms, geysers are transitory because the precise conditions that create them are never maintained for long. Often the flow of groundwater will slacken, so the geyser erupts at longer and longer intervals (in some places, geysers are "encouraged" to perform by dosing them with water). Quite often, a geyser's subterranean chamber becomes eroded by its superheated water and eventually collapses. But even if a geyser escapes both these changes, its final death is assured, after centuries or millennia, by loss of heat from the cooling rocks beneath.

Pohutu Geyser, at Rotorua in New Zealand, erupts in a cloud of superheated steam (right). Like many geysers, its scalding underground water dissolves large amounts of minerals from the surrounding rock. When the water reaches the surface and cools, it rapidly sheds its burden, creating strange pans and shelves as it flows away.

The geyser of Strokkur, in southwest Iceland, erupts at roughly nine-minute intervals. From a quiet steam-covered pool, Strokkur – "the churn" – swells into a dramatic dome of scalding water (left). On eruption, a shaft of steam and boiling water shoots up to 100 feet (30 m) and is visible 3 miles (5 km) away.

The history of rock

Rock and stone are proverbial symbols for anything strong, steadfast and inert. But as a worn gravestone shows, everything on the Earth is in a state of change. Even though change may be infinitely slow, not even the hardest rock escapes the general rule.

All rocks are composed of minerals, which are solid substances with a specific chemical "recipe". Minerals, in turn, are made up of elements – different types of matter each composed of just one kind of atom. A few elements, such as gold and sulphur, sometimes exist in a pure or "native" state in the Earth's crust, but far more are locked together in chemical combinations.

Some of these chemical combinations are quite simple. Rock salt, for example, is made up of equal numbers of sodium and chlorine ions (an ion is an atom that is electrically charged), linked together in an orderly way. Quartz, one of the commonest minerals on Earth, consists of silicon and oxygen ions. In this case, the ratio between the two elements is not one-to-one, but one-to-two. Other minerals can be far more complex. Serpentine, a green mineral used in jewellery, contains four elements – magnesium, silicon, oxygen and hydrogen – with three ions of magnesium for

every two of silicon and every nine of oxygen.

A perfect crystal of rock salt or quartz is a mineral in its purest form. But most rocks are not like this. A close look at a piece of polished granite shows that it contains countless small crystals of different minerals, fitting together like a random three-dimensional jigsaw. Each crystal is held together by strong internal forces, and it is locked into its neighbours by virtue of its irregular shape. This helps to give granite its strength: because there are no lines of weakness, it is very hard to break.

Granite is a familiar example of an igneous rock, one that has been formed from magma. The crystals in granite and other igneous rock develop as the magma cools, and the slower the rate of cooling, the bigger the crystals. Magma usually solidifies before it reaches the surface. However, over millions of years, the rock it produces can be lifted up and exposed to ice, wind and water. Despite its great strength, igneous rock cannot withstand this treatment for ever, and it slowly disintegrates, forming particles that are dissolved or swept away to create banks of sediment.

If the sediment becomes deep enough, physical and chemical changes transform it

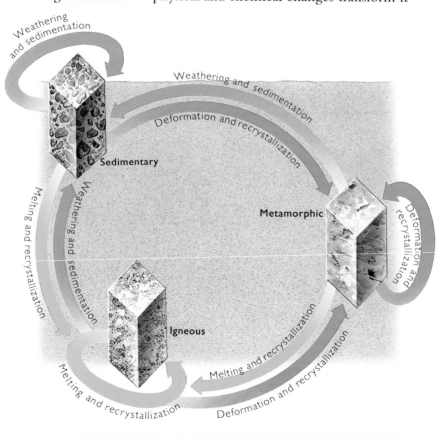

Natural cycles are a part of daily life. Seeds germinate to produce plants, and plants in turn produce seeds. Heated water creates steam, and cooled steam creates water. The same is true of the three classes of rock. By various processes igneous, sedimentary and metamorphic rock are transformed from one to another, and into different forms of themselves.

The speed of different parts of the cycle varies. A single layer of sedimentary rock may take thousands of years to build up and centuries to erode. But when magma comes into contact with existing rock, its abundant energy can trigger changes that occur in days or even minutes.

The Devil's Marbles in Australia's Northern Territory. The boulders are probably the result of chemical weathering of igneous rock through cracks. As susceptible minerals along the joints decompose, the joints widen, breaking the rock into irregular blocks.

from a soft slurry into a layered sedimentary rock. This transformation is just one of many that take place in and on the Earth's crust; it is a stage in an endless cycle that involves the whole of the Earth's surface.

Heat energy from nuclear reactions, both in the interior of the Earth and in the Sun, powers the cycle. The Earth's heat produces upwelling magma which forms igneous rocks and transforms existing rocks. It also drives the movement of tectonic plates, which lift rocks upward to form mountains, and which plunges them into the crust where they are melted. Heat from the Sun produces rain, snow and wind, which wear away rocks at the surface. The Earth's rocks are thus exposed and buried, heated and cooled, squeezed and scattered, changing their character with every step.

Hot rocks

The surface of our planet is deceptive. Much of it is covered by rocks formed from sediments, or by rocks that have been fashioned from them. But this material makes up only 5 percent of the Earth's crust. The rest is igneous rock – rock that has been forged by heat.

Every year, the heat of the Earth's interior drives several cubic miles of liquid magma upward on a journey through the crust. The crust is thinnest beneath the oceans, and here rising magma often stays hot enough to arrive at the surface as a liquid. When it meets the ocean floor, it oozes out over the seabed, turning solid as its heat is rapidly quenched by the near-freezing water. Beneath the sediment on the floor of the oceans lies a flat or crumpled sheet of basalt, an extrusive igneous rock. Hard, black and finely grained, basalt is probably the most common rock on Earth.

Magma that rises upward through the continents has a much longer journey to complete. Some of it remains liquid and pours out through volcanoes or fissures, but much more slows down as its heat is gradually drained away, and eventually comes to a halt.

The result is intrusive igneous rock which has been frozen solid on its way to the surface.

Geologists have borrowed the Roman god of the underworld to provide a name for these hidden masses of frozen magma. A pluton is a subterranean intruder from deep in the crust, a body of igneous rock – often granite – that has squeezed its way into the older rock around it.

Like a jelly setting in a mould, solidifying igneous rock takes on the shape of its surroundings. Some of these shapes are strange indeed. They include immense, irregular lumps and steeply sloping sheets that can be inches or miles across. Where the magma has squeezed its way between layers of existing rock, it often forms flat-sided shelves. If it can lift the layers above it, it bows the rock into a curve, creating a shape like a giant lens.

Intrusive rock is normally hidden because most of it solidifies within the crust. But as uplift and erosion refashion the surface, the rock is often exposed. The granite tors of Dartmoor in southwest England are ancient plutons that have been exposed and eroded.

Molten magma makes its way upward by engulfing fragments of the surrounding rock, or by seeping through its faults and joints. In some places, the intruding magma follows the existing layers of rock, forming an extra layer like the filling in a sandwich. In others, it cuts straight across the layers, and pushes them apart.

The largest kind of volcanic intrusion is a batholith – a domed, irregular mass formed deep beneath the surface. A boss is a cylindrical intrusion with steeply inclined sides, and it is usually smaller than a batholith.

Magma flowing upward into a volcano produces a cone-shaped mountain of lava, which may later be eroded by the elements to expose a volcanic neck or plug. By contrast, magma seeping upward through vertical fractures forms dykes – flat, upended sheets of rock. In some parts of the world, dykes are crammed together in "swarms", where the igneous rock has followed parallel fissures on its way upward.

A sill has the same shape as a dyke, but follows the existing rock layers rather than cutting across them. Lopoliths, laccoliths and phacoliths are all intrusions that follow the existing strata.

The Devil's Tower in Wyoming is a volcanic plug – a mountain formed by rock that was once at the core of a volcano. The ash that makes up a volcanic cone is quickly eroded when eruptions cease, but solid magma is better at withstanding the test of time. The parallel columns of the tower were created because the magma shrank and cracked as it cooled, taking on crystal-like shapes.

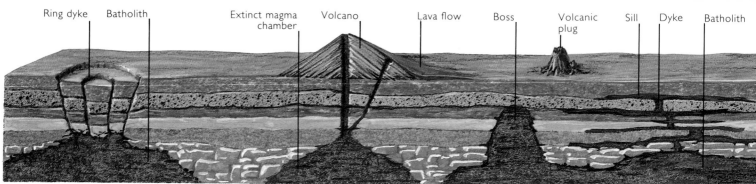

Ring dyke   Batholith   Extinct magma chamber   Volcano   Lava flow   Boss   Volcanic plug   Sill   Dyke   Batholith

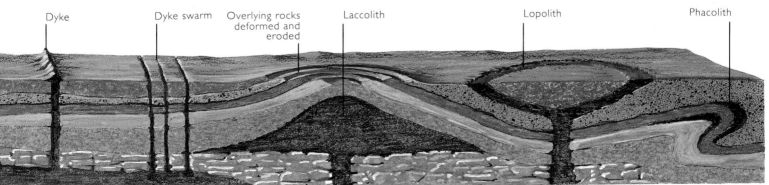

Dyke | Dyke swarm | Overlying rocks deformed and eroded | Laccolith | Lopolith | Phacolith

Layered rocks

The exposed surface of the Earth is under constant attack. The agents of erosion – ice, wind and water – scour and shatter rocks, and carry off the fragments. Soft rock quickly succumbs to this onslaught, but even a rock as hard as granite cannot withstand it for long.

The ultimate destination of most of these fragments is the sea, where they finally sink to the bottom, to build up into deep banks of sediment. Sedimentation is often a cyclical process, waxing and waning as the seasons change. Each year produces a visible band of sediment, called a varve.

As layer smothers layer, the pressure beneath increases, and the deepest rock fragments become squashed closer and closer together. At the same time, they are often saturated by water that is rich in dissolved minerals, such as silica. These minerals often crystallize in the spaces between the particles, and once this

Baked by the harsh desert sun, a crust of salt shows where flood waters once stood in California's Death Valley.

Salt crystals are a special form of sediment – one that is formed by chemical deposition. This sort of sediment formation can happen when water evaporates and when it is still liquid. If the concentration of a dissolved mineral becomes great enough to exceed saturation point – as sometimes happens in enclosed seas – the mineral begins to take on solid form, and settles out of the water.

Underground domes of rock salt, which can be several miles high, are formed by sea salt that has been deposited in this way.

Rock-forming sediment does not only come from land. In the sea, pinhead-sized skeletons of tiny animals trickle down from surface waters in billions, to form deep oozes hundreds of feet thick.

The skeletons here are those of foraminifera, single-celled animals that have a perforated body shell of calcium carbonate. When compressed and cemented, these produce a fine-grained "organic" limestone. Limestone formed by the remains of living things has many different textures. At one extreme is chalk, a smooth, powdery rock made of very small skeletons, and at the other coquina, a coarse, brittle rock made of large shells cemented together.

"Inorganic" limestone is created from calcium carbonate shed directly by sea water. The calcium carbonate often accumulates in small spheres, called oolites, which produce distinctive oolitic limestone.

"cementation" is complete, the particles are glued together to make solid rock.

Sedimentary rocks can be aged by their layers. A shallow roadside cutting can reveal thousands of years of sedimentary history at a glance, while on coastal cliffs the sea may expose the remains of sediment that took 100,000 years to accumulate. In the Grand Canyon, over three billion years of geological history are on show in a deep "slice" a mile (1.6 km) deep through the Earth's crust. Over time, the record in rock has often been erased by erosion or masked by geological movements, which have folded and distorted layers.

Transformed rocks

For more than 2,000 years, the quarries at Carrara in northern Italy have yielded a pale and lustrous marble that is one of the most famous rocks in the world, not least because it was Michelangelo's favourite medium.

Marble is an example of a metamorphic rock, one that has undergone a profound change through the effects of heat or pressure. A single precursor rock can give rise to many different kinds of metamorphic rock, depending on local conditions in different parts of the crust.

Metamorphism takes place when an igneous or sedimentary rock becomes distorted or heated, and the changes occur when the rock is still largely solid. One of the most common causes of metamorphism is the collision of tectonic plates. This bends and reshapes rocks as mountains are formed, throwing once horizontal layers into complex curves.

Tectonic collisions can also bury rock, subjecting it to increased heat and pressure which again trigger changes in the rock's make-up. Another cause of change is heat

from nearby igneous rock, which often forces its way upward from deep in the crust.

When rock is baked by heat and pressure, the mineral crystals break down and realign, giving it a new texture. If the heat and pressure are moderate, the changes are slight, and the result is a low grade metamorphic rock. But if the baking process brings the rock almost to the point of melting, a greater transformation occurs. The mineral crystals grow much larger, and the result is a high grade metamorphic rock, which is hard, crystalline and brittle.

Shale and schist are two rocks that look different, but which are directly linked in this way. During moderate baking – at about 480°F (250°C) and a pressure of 2,000 atmospheres – shale is transformed into slate, a rock that splits easily into flat sheets. If the temperature is increased to about 750°F (400°C), slate is transformed into phyllite, a harder rock with larger mineral crystals that give it a rough texture. And if phyllite is baked even more fiercely, the result is schist, a sparkling rock that is common in mountains.

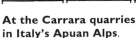

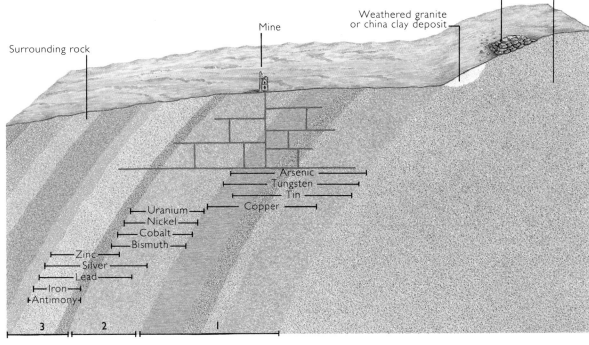

At the Carrara quarries in Italy's Apuan Alps, marble is sawn into blocks to be transported all over the world. Marble is formed from limestone by heat. As the limestone warms up, its layered structure disappears and is replaced by a jigsaw of calcite crystals. This gives it a granular texture and makes it harder.

If the crystals are relatively pure the rock is a pearly white colour. If the limestone contains impurities, these produce the beautiful streaks of red, grey or green which make some marbles so distinctive.

The strangely sculpted tors of Dartmoor, in southwest England, are the surface signs of a huge body of granite that lies hidden below ground. The granite formed from molten magma.

Heat released by the setting granite permeated the surrounding rocks, transforming them through contact metamorphism. The highest-grade metamorphic rocks lie closest to the granite, where the heat was most intense, while farther away, the metamorphism is less marked.

The granite also produced superheated water, laden with minerals, that flowed through cracks in the rock. As the water cooled, it shed its mineral burden, depositing ores of metals such as tin, copper, lead and iron.

Each ore was shed where the temperature and pressure of the surrounding rock fell below a certain level. As a result, the ores are found in different bands that encircle the granite mass.

The diagram shows the zones in which ores are likely to occur; however in reality they are not all found together.

1 **Hypothermal zone**
575–925°F (300–500°C)

2 **Mesothermal zone**
400–575°F (200–300°C)

3 **Epithermal zone**
125–400°F (50–200°C)

From rocks to jewels

Diamonds and coal are two materials formed by identical carbon atoms arranged in different ways. Although both are useful, weight for weight, one is about five million times more valuable than the other. The reason for this enormous disparity lies in three special qualities possessed by all gems – durability, beauty and, above all, rarity.

Coal is present in vast quantities in the Earth's crust, and much of it is easily accessible. However, it easily chips and splits once brought to the surface. By contrast, diamonds are made of the hardest material known. Like rubies, sapphires and emeralds, diamonds look unremarkable in their natural state, but once they have been cut, they become objects of extraordinary beauty. The flat faces of a cut stone are carefully angled so that the jewel bends and reflects nearly all the light that falls on it.

To add to this allure, diamonds are rare. In the most productive mines, in the kimberlite rock of southern Africa, over 500 tons of ore have to be mined and processed to yield an ounce (30 g) of diamond. Most of this will be made up by small diamonds that have industrial value only, leaving just a small proportion that can be made into jewels.

Gemstones need precise conditions in which to form. Most gems are made of silica, or of metal silicates and oxides (diamonds being an exception). If the temperature and pressure are right, and there is a sufficient supply of one of these minerals, it may form crystals. But if it gets too hot, a crystal may dissolve, and if the supply of minerals fails, the crystal's growth will cease when it is still microscopic in size.

The right conditions for crystal growth are usually found only at considerable depths in the Earth's crust. This is where most gems form, and they would remain here, were it not for the crust's constant upheavals.

Uplifted rock or rising magma brings gems much closer to the surface, where they can be mined, or where erosion can release them from their parent rock. Because most gems are hard, they withstand erosion better than other minerals. As the rock breaks down, they are swept into rivers, to sink under their weight and await the bulldozer or the prospector's pan.

1 Aquamarine
A variety of the mineral beryl. Bluish-green colour due to impurities of iron. Formed in cavities in the metamorphic zone around a granite intrusion. World's largest source is Brazil.

2 Heliodor
Yellow variety of the mineral beryl. Traces of iron give its colour. Formed in cavities around a granite intrusion.

3 Morganite
A variety of the mineral beryl. Local impurities of manganese produce its pink colouring. Like aquamarine and heliodor, it is formed in granites and pegmatites – very coarse-grained igneous rocks. The best source is Brazil.

4 Sapphire
The name given to all varieties of the corundum mineral with the exception of rubies, which are coloured only by chromium. The colours vary from deep blue to purple, pink, orange and green depending on the amounts of iron, titanium and chromium present. Formed deep underground as a result of metamorphic reactions in shales and limestones. Sri Lanka is the main source.

5 Emerald
A variety of the mineral beryl. Traces of chromium produce its distinctive green colour. The gem crystallizes in cavities within cooling granite. Colombia mines the world's finest gems; other sources include Zambia, Zimbabwe and Pakistan.

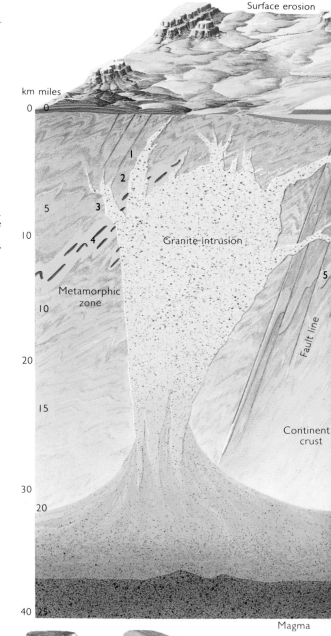

Surface erosion

km miles
0   0

5

10

10

20

15

30

20

40   25

Granite intrusion

Metamorphic zone

Fault line

Continent crust

Magma

Jadeite

Aquamarine   Amazonstone

Morganite

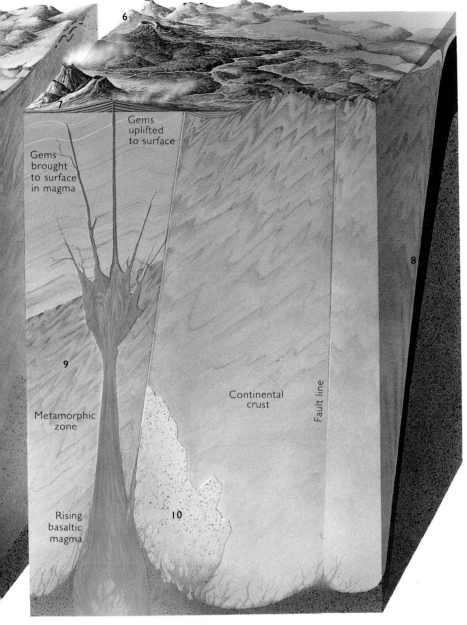

Gems
uplifted
to surface

Gems
brought
to surface
in magma

6

7

9

Metamorphic
zone

Rising
basaltic
magma

10

Continental
crust

Fault line

8

6 Amethyst
A form of the mineral quartz, one of Earth's commonest. Quartz occurs in many igneous and metamorphic rocks. Iron impurities produce the amethyst's vivid purple colour. Formed as large crystals in cavities in the cooling rock.

7 Amazonstone
Part of the feldspar mineral group. Impurities of lead and water produce its greenish-blue colour. Formed in cavities as volcanic rocks cool. First found near the Amazon River in South America.

8 Jadeite
Semiprecious mineral of the silicate group. Formed under intense heat and pressure at a subduction zone, it is later uplifted to the surface. It is mined in Burma, China, Tibet, Guatemala and Japan.

9 Ruby
A variety of the mineral corundum, consisting of aluminium oxide. Chromium impurities produce the rich red colour. Formed in metamorphic rocks deep underground where the intense heat alters the aluminium-rich sediments. Burma, Kenya, Zimbabwe and Tanzania mine rubies.

10 Peridot
The gem variety of the mineral olivine, found in basalts and other ultrabasic rocks. The shade of green is determined by the amount of iron present. It is mined in Burma, Arizona, U.S., and Norway.

Most gemstones are a by-product of metamorphism beneath the Earth's surface, in which heat or pressure transforms one form of rock into another. During this process, scarce minerals are sometimes concentrated in pockets of rock or in intensely hot water. As the temperature falls, the minerals crystallize out of solution. The growth of gem crystals takes place slowly, and often at great depths.

Diamonds and some kinds of garnet form where the crust meets the mantle, more than 60 miles (100 km) beneath the surface, where the temperature is above 1,800°F (1,000°C). Rubies and sapphires develop from the metamorphism of deep layers of sedimentary rock, while topaz, aquamarine and a number of other gems develop from minerals concentrated in pegmatite, a rock formed by cooling magma. Gems that occur near the surface, such as turquoise, opal, amethyst and agate are created by the evaporation of mineral-rich fluids, such as silica-bearing water, or by the breakdown of volcanic rocks.

When gem crystals develop, they sometimes trap impurities that affect their colour and also their value. Pure corundum, for example, is a colourless mineral made of aluminium oxide. But if chromium atoms are locked in the corundum crystal as it grows, the result is a ruby — a highly sought-after and precious jewel. In a similar way, pure diamond has no colour. Valuable "blue" diamonds are created when atoms of boron are sealed into the crystals' carbon lattice.

Ruby

Emerald

Peridot

Sapphire

Heliodor

Amethyst

Falling stars

In 1609, the Italian physicist Galileo used his telescope to study the Moon, and became the first person to see the round depressions that are scattered over its sterile surface. These ragged circles are now known to be craters – the legacy of a massive onslaught from space.

The Moon's craters were formed by meteors – fragments of matter that ploughed into it after a journey through the solar system. Evidence of these impacts can be seen on Mars and the moons of other planets; on Mercury, a single giant impact created the Caloris Basin, which measures 800 miles (1,300 km) across.

Most of these craters were created during the Great Bombardment, a period about four billion years ago when the solar system was cluttered with debris orbiting the Sun. As today's planets took shape, they absorbed anything that came close enough to be influenced by their gravity. Since then, the planets have "mopped up" much of the solar system's debris, but a great deal still remains. The sudden streak of a shooting star in the night sky shows that like all the planets, Earth is still gathering in its share.

Every year, the Earth attracts more than a million tons of new material from space. But because it has an atmosphere, most of these pieces of matter are vaporized before they reach the ground. On rare occasions, they survive their descent and are called meteorites.

So far, about 2,000 meteorites have been found, and once their blackened outer surfaces have been cut away, they prove to be of three distinct types – made either of iron, of iron and minerals, or of a stony material. Each type corresponds to a layer in the Earth's interior. This suggests that many meteors formed after the break-up of planetesimals – bodies formed during the solar system's birth which were large enough to develop a layered structure, like the one that the Earth now has.

The Great Bombardment ended when the Earth was still in its infancy. But this does not mean that giant meteors are entirely a thing of the past. Statistically, it is only a matter of time before another major impact occurs and the Earth takes aboard another of these relics of the solar system's past.

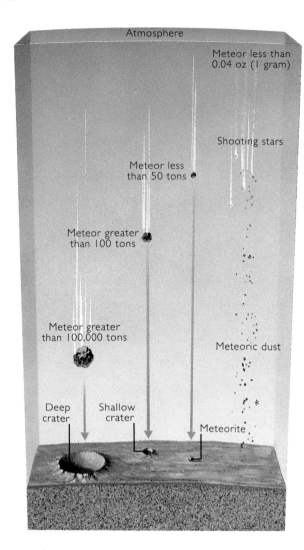

The ultimate fate of a meteor travelling toward Earth depends on its size. Giant meteors, such as the one that produced Meteor Crater in Arizona, punch their way through the atmosphere, but are often completely obliterated during a devastating impact with the ground.

Meteorites that are of intermediate size are slowed by their passage through the atmosphere. Although much of their material is vaporized by friction, a core may remain intact and reach the surface, pitted and sculpted by its passage through the air. It is then called a meteorite. The heaviest core yet recovered, from Namibia, weighs more than 50 tons.

Smaller meteorites frequently break up into fragments during their descent, producing shooting stars, intense streaks of light that flash across the sky. The smallest particles, specks of matter smaller than grains of dust, slowly settle through the atmosphere and arrive intact at the surface weeks or even years later.

Although invisible, these micro-meteorites can be identified in the sediment of the sea floor or in ice. Meteor showers, which are visible at certain times of the year, are produced when the Earth moves through bands of debris orbiting the Sun.

Arizona's Meteor Crater is one of about 200 craters that have been identified on Earth. The object that made this giant crater was probably about 100 feet (30 m) across when it reached the ground, and by then it would have been "slowed" to a speed of about 30,000 mph (50,000 km/h).

As it struck the ground, its explosive impact punched out a crater that is today more than 650 feet (300 m) deep and ³/₄ mile (1.2 km) across. The shock wave produced by the impact would have killed animal life over a wide area, and the dust generated would have travelled around the world.

Two-thirds of the meteors reaching the Earth hit the sea, leaving no trace. The ones that do hit land produce craters, but they do so on a surface that is constantly, but imperceptibly, changing. Unlike on the Moon, weathering and erosion gradually obliterate the outlines of craters, so they become indistinct and eventually invisible. Many thousands of years from now, the Arizona crater itself will have disappeared.

Life turned to stone

Without fossils, we would know very little about the story of life on Earth. Fossils are the buried remains of living things, sandwiched in layers of sedimentary rock. All forms of life gradually change or evolve with the passing of time, and the Earth's fossils reflect these changes. Together they are like a mounting pile of snapshots in a vast and growing family album.

Like any photographic collection, the fossil record has its gaps, and in places it is given to favouritism of a most blatant kind. Some fossil species – such as the familiar ammonites and trilobites – appear in vast numbers, while millions of equally successful families hardly feature at all. These so-called missing species eighter had no hard parts from which fossils could form, or they lived in places where the process of fossilization rarely occurred. Even where fossils have formed, they have not always survived the test of time. In some places, fossils have been built up over millions of years, only to be erased by metamorphism, which transforms sedimentary rocks and destroys the fossils they contain.

Despite these shortcomings, the fossil record gives a unique insight into the world long ago. The giant fossilized bones of huge creatures such as dinosaurs and early mammals have a stunning impact because of their sheer scale,

but to a palaeontologist – someone who studies fossils – the fine structure of smaller fossils often proves to be more revealing.

A fossil pollen grain, looked at under a microscope, will show a pattern of tiny bumps and pits that identifies it as surely as a fingerprint. The existence of the pollen grain proves the existence of its parent plant, and so gives important clues about climate. In the same way, a fossil shell will often show annual growth rings, establishing how old its owner was when it died. These rings will also show how fast the animal grew, and therefore what the climate and food supply were like when it was still alive.

As well as revealing information about individual species, the fosil record charts the history of the living world as a whole. Over millions of years, it shows that life has experienced periods of rapid growth, but also some startling setbacks, during which the number of species alive seems to have plunged. These drops are known as mass extinctions and their causes are a subject of hot debate. There is some evidence that the famous mass extinction at the end of the Cretaceous period, which involved not only the dinosaurs but the majority of species then in existence, was triggered by a meteor, but the causes of many other extinctions are unknown.

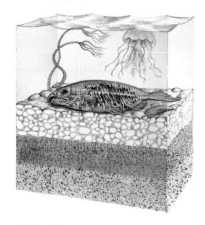 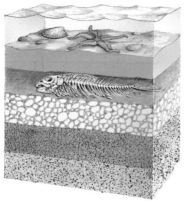  

190 million years ago
Dapedium dies and sinks to the seabed. The fleshy tissues are recycled by bacteria, but any hard parts, such as teeth and bones, remain.

130 million years ago
Layers of sediment settle on the bones, which undergo a chemical change. The original minerals are dissolved and replaced by ones from sea water.

50 million years ago
The fossil skeleton, now on dry land, is compressed and distorted as the sediment layers are first compacted, then uplifted and tilted by land movement.

Today
Long after the fossil formed, slow geological movements can bring it back to the surface. Here, erosion may release the fossil from its rocky tomb.

A relic from early Mesozoic seas, *Dapedium punctatus* grew to about 1 foot (30 cm) long. It fossilized well because it lived near the shore, where there was a good supply of sediment to smother it after death.

Like many other fish of this period, it also had an armour-plating of thick, hard scales which preserved the outline of its body one its soft tissues had been broken down. Fossils such as this one caused great confusion among early geologists, because they were often found far away from the present-day sea, even near mountain tops.

Until the 17th century, there was debate about whether fossils were indeed organic remains or whether they were objects that had always been inanimate, like rocks and minerals. By the 19th century, it was accepted that fossils were remnants of species of plants and animals that were now largely extinct. Charles Darwin used changes seen in the fossil record as evidence to support the theory of evolution.

Bountiful Earth

When Edwin Drake drilled the first oil well at Titusville, Pennsylvania, in August 1859, he began a revolution in the way we live. For nearly a century and a half, oil has provided us with a cheap and convenient source of energy. It can be pumped, piped, stored in tanks and distilled into a whole range of products. It burns easily, and it produces intense heat without leaving any solid waste. Oil is an almost perfect fuel; unfortunately, it is also one that is rapidly running out.

Like coal, oil is a fossil fuel, an organic substance that has been transformed by geological forces over millions of years. As yet, geologists are unsure about how oil formed, but the starting point was almost certainly microscopic plant and animal remains that fell on the seabed from the sunlit waters above.

These organic "leftovers" became smothered by layers of sediment. Once buried, they were partly decomposed by bacteria, which altered their chemical composition to produce hydrocarbons – energy-rich compounds of hydrogen and carbon. Some had long chains of carbon atoms, giving them a thick, tarry consistency. Others had chains only one or two atoms long, and these substances formed natural gas, trapped deep underground. Millions of years later, when the hydrocarbon "soup" is brought to the surface to be refined and then burned, the stored energy that binds the carbon and hydrogen atoms is released.

It is easy to imagine that fossil fuels formed only in the past, but they are still being formed today, although not nearly as fast as they are being used. The swamps of places such as the southeastern United States are probably producing the coal of the future, and organic remains are still settling on the seabed, forming the raw materials of oil.

Coal has been exploited for hundreds of years, and with reserves still running into billions of tons, it is set to last several centuries more. However, the situation with oil is very different. Worldwide, oil is consumed at the rate of 1,000 barrels every second, and the figure continues to increase. At the present rate of consumption, most of the readily accessible oil will have been tapped by the year 2020.

Oil can be produced from coal, but another possibility is to distil it from oil shale. This soft, sedimentary rock, which is found in many parts of the world, contains thin films of oil sandwiched between layers of rock. The oil can make up to 20 percent of the total weight, and it can be driven off by heat. Oil shale deposits may represent the largest deposits of fossil fuels in the world. However, in most oil shale, the proportion of oil is meagre, and may be too little for it to be economically viable.

Coal, a sedimentary rock, forms from the remains of swampy tropical forests, similar to the peat bogs of today. The lack of oxygen here inhibits the natural processes of decay, so plant remains accumulate as peat (1).

Over millions of years, overlying rock compacts the peat into lignite or brown coal (2). Further pressure and heat burn off impurities creating soft, black bituminous coal (3), the commonest form, used to make coke for industry. Metamorphism produces hard, shiny anthracite (4), almost pure carbon. Although it is the most effective fuel, it is rare and costly to mine.

Swampy forest — Dead vegetation forms peat — Peat slowly transformed into lignite — Further compression forms bituminous coal — Anthracite coal seam finally forms

1   2   3   4

Reservoir rock

Oil and gas migrate upward

Oil and gas trapped

Non-porous rock

Marine organisms decay

Oil and natural gas form

Fault

1

2

3

4

Oil forms in undersea sediments from the remains of minute organisms that live in warm shallow seas. Layers of these phytoplankton sink to the stagnant, oxygen-free seabed and are soon buried (1). Millions of years of heat and pressure deep below ground transform this partly decayed organic matter into oil and gas (2).

Porous rock allows the oil to migrate up toward the surface (3). Small amounts are trapped by non-porous rocks in reservoirs, or oil pools, similar to water aquifers. These can be tapped by oil wells (4), like this one in the North Sea (above).

73

SHAPING
THE LAND

Shaping the land

The forces of rain, rivers, ice and wind are relentlessly wearing away the mountains, valleys, plains and plateaus of the planet. They attack the land, breaking it down mechanically into smaller and smaller pieces or chemically dissolving it into solution. The resulting sands, gravels, boulders

and solutes travel long distances and are either used as building blocks for new landscapes or are washed into the oceans. Sediments accumulate gradually on the sea floor and after many millions of years they are compressed, contorted and uplifted to form mountain ranges and a new cycle of destruction and construction begins.

The extent to which a landscape is modified by weathering depends not only on the power of the individual processes but also on the length of time over which they operate, and the resistance that is encountered from soil and rock. Bare soil slopes are more readily eroded than vegetated slopes whose plants bind the soil particles together. Massive rocks like granite resist attack more effectively than soft sandstones and clays.

The power of weathering often varies with the seasons. Occasionally short bursts of torrential rain, river floods, glacial surges and hurricane-force winds effect dramatic landscape changes which would otherwise have taken decades or centuries to achieve. But generally weathering is a leisurely, almost imperceptible process.

Many of today's most striking vistas in the northern continents are relics of a prolonged and bitterly cold period when ice sheets, glaciers and frost action dominated. Around 20,000 years ago mountain glaciers sculpted cirques, as well as hanging valleys and U-shaped valleys out of the land. Ice sheets bulldozed the eroded sand, gravel and boulders into huge mounds while summer torrents flowing from the melting ice created vast outwash plains and long winding ridges.

Today the harsh outlines of many of these features have been softened by vegetation but they have not yet been completely obliterated by the subsequent effects of rain and rivers. Similarly, some regions that are now wet and tropical still harbour the remains of desert sand dunes, while subtropical deserts, such as the Sahara, display landscape features that indicate that vegetation once flourished here when it was shaped by a wetter climate.

One of the most exciting challenges for landscape detectives, known more scientifically as geomorphologists, is to unravel the complex nature of the actions that have created the diversity of present landscapes. This involves not simply looking at the shape of the features but also viewing them in the context of the setting, and analysing any scars they bear as well as their component sediments.

Sediments can often provide clues as to when a landform was created, how long it took and the prevailing climate. Those laid down in lakes, rivers and swamps often contain preserved pollen and the remains of beetles, snails and animal bones. Identifying the different species of pollen and animals that are preserved in each layer can reveal the

Sculpted and scoured by the elements of wind and water into strange and surreal shapes, these desert rocks (previous page) bear mute witness to their origin millions of years ago on the bed of the mighty Tethys Ocean.

changing environmental history of an area over thousands of years.

Rain and rivers are the dominant forces at work on many of the planet's landscapes. From the moment that raindrops strike the bare earth, especially during heavy rainstorms, landscape destruction begins. Dislodged soil particles are carried overland to the rivers and washed away as suspended sediments. Alternatively, water seeps into the ground where it works its way slowly down, dissolving minerals from the soil and rock as it goes, before eventually reaching the river. In limestone rocks, where chemical weathering is especially effective, fascinating and beautiful underground landscapes of caverns, lakes, stalactites and stalagmites are carved.

Athough a river may look clear, it is actually eroding the landscape before our eyes. Half of the material carried is in the form of invisible solutes and the river adds further suspended sediments to its load by undercutting the river banks and valley sides which then collapse. During floods the turbulent river water becomes so powerful it can move boulders the size of a minibus along its bed.

A river's power to erode and destroy is evident the length of its course, but it also has the ability to construct landscape features. Where it slows down to a meander it deposits large amounts of sediment and creates flood plains, levees and even islands. Eventually, the sediment reaches the sea where it either develops a delta or is swept away by ocean currents before settling on the sea floor.

In cold climates it is not only large-scale features such as ice sheets and glaciers that shape the land. Water penetrates cracks and crevices of rocks by day and freezes at night, expanding its volume by up to 10 percent. Repeated freeze-thaw cycles loosen grains, shatter rocks and liberate blocks which, under the force of gravity, tumble down to the foot of slopes as screes.

In subpolar regions, soils are often frozen solid throughout the winter in permafrost, or tundra, thawing at the surface only briefly in the summer, if at all. Over centuries this seasonal cycle opens up cracks in the ground and causes the soil to heave up into mounds, often shifting stones in the soil to the mound edges.

Winds play only a minor role in shaping landscapes in moist climates, but in arid regions – one-seventh of the planet is desert – they are the dominant weathering agent. Wind-blown sand scours rocks and sculpts them into smooth pavements, mushroom-shaped pedestals and commanding pillars as well as heaping up migrating sand dunes in ever changing forms.

Natural forces will continue to alter the face of the Earth but their effects are being accentuated by human actions. The removal of vegetation and cutting roads in steep hillsides encourages soil erosion; overgrazing or intensive cultivation of crops is transforming vast areas into deserts; pollution is creating acid rain which hastens chemical weathering. There is now an urgent need for us to improve our understanding of how natural forces shape the landscape, and to use this vital knowledge to limit the environmental degradation we are causing.

Rivers at work

Not only do rivers carry excess water from the land to the sea, but they also have the power to transform a landscape over time with their ability to erode, transport sediment and then deposit it. In fact, rivers account for more surface features than any other natural agent and they are found in virtually every part of the world. All rivers produce valleys and it is within these valleys that the array of erosional and depositional features is etched.

Rivers are the vital link in the water cycle between the atmosphere, from which precipitation falls, and the oceans, from which water evaporates. Each river is part of a much larger network of interconnected streams and tributaries organized in a drainage system. Each system is, in turn, dominated by a main channel, joined by tributaries which flow from all parts of the drained area, starting near the watershed. Close to the sea a river may divide into a network of channels spread over a delta.

A river can begin its life as a spring, as meltwater from a glacier or, most commonly, as overland flow. Before overland flow can begin, rainwater must sink into the ground until it becomes saturated. Once the ground has soaked up all it can, any hollows fill up, and soon minute, threadlike channels, or rills, carry water away. Eventually these small channels join to become steep-sided gullies and, ultimately, true stream channels.

When these are joined by others, a river develops. The amount of water in a river depends to a large extent on the amount of precipitation that falls in the river's catchment area and this changes with the seasons. Generally, in temperate areas the flow is greatest during the winter because of high rainfall. But in glacial regions rivers are swelled by spring meltwaters. In arid areas rivers only flow during storms.

The amount of water held in the world's rivers at any given instant is only a fraction of that stored in lakes, in the groundwater system and in the sea. But rivers are capable of moving huge volumes of water, sometimes at high speed, and it is the constant motion of water that enables rivers to erode and transport vast quantities of material.

Source zone

The source of a river is often in an upland mountain region, where precipitation levels are high. Where surface runoff collects into small drainage channels that take it downslope, a river system begins. In glaciated areas a lake may provide a temporary store for the water, which then emerges as an outlet stream.

Many tributaries enter the main channel running off the higher land. The dominant process is erosion, the fast-flowing stream incising deep into the rocks as it seeks the shortest route to the sea.

Waterfalls are created where a difference in height forces the river to cascade over a vertical drop. This falling wall of water erodes a plunge pool at its base. Rapids are places where the flow is disrupted to create fast turbulent waters. A gorge is formed when the slow uplift of land forces a river to continue eroding its channel, cutting steep, almost vertical walls.

Transition zone

Away from the wet uplands, rivers begin to flow at a more sedate pace and their power to erode is lessened. Here the transport of sediment dominates as input from further erosion is roughly balanced by sediment lost to deposition.

A braided channel arises along the steeper reaches, where the river is carrying a large quantity of coarser material. The flow is diverted around islands and bars of deposited sediment, creating a series of interwoven subchannels.

A meandering river creates broad, looping bends. Steep slopes produced by undercutting of the outer banks form bluffs where the flow is greatest. On the inside bend of the meander, sediment is deposited as point bars or scroll bars. River terraces may form

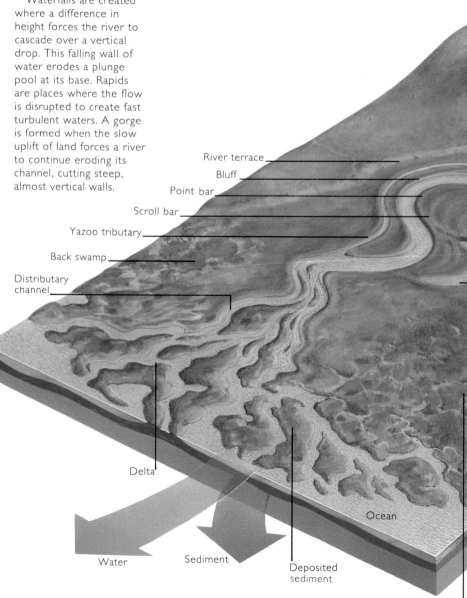

River terrace

Bluff

Point bar

Scroll bar

Yazoo tributary

Back swamp

Distributary channel

Delta

Water

Sediment

Deposited sediment

Ocean

Salt marsh

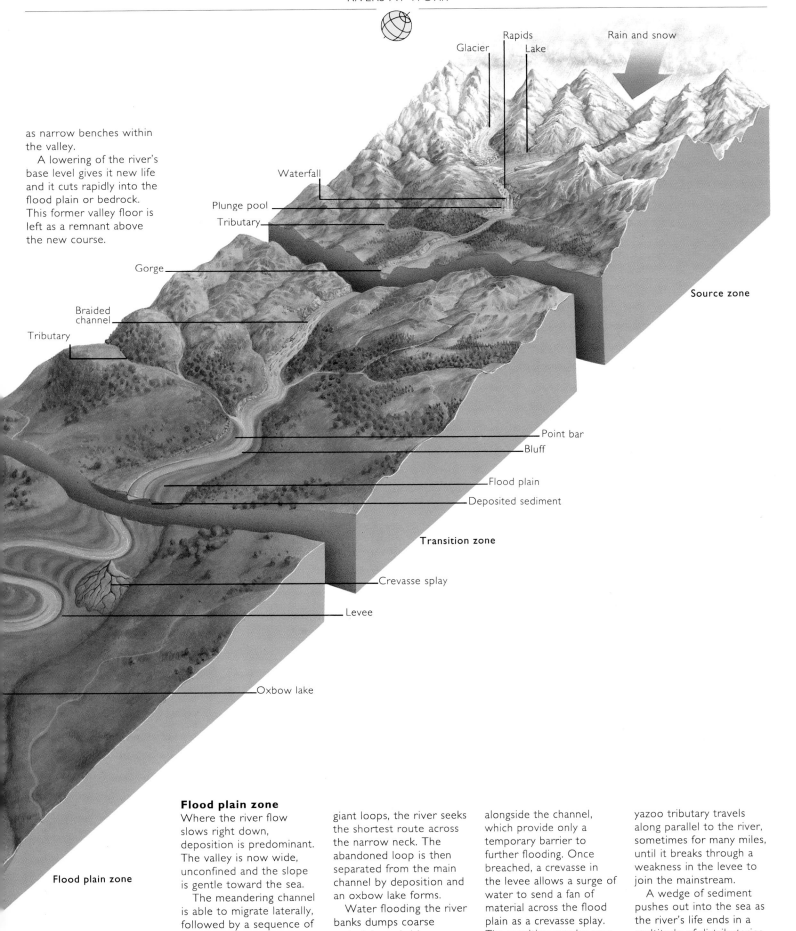

as narrow benches within the valley.

A lowering of the river's base level gives it new life and it cuts rapidly into the flood plain or bedrock. This former valley floor is left as a remnant above the new course.

Waterfall

Plunge pool

Tributary

Gorge

Braided channel

Tributary

Rain and snow

Glacier

Rapids

Lake

Source zone

Point bar

Bluff

Flood plain

Deposited sediment

Transition zone

Crevasse splay

Levee

Oxbow lake

Flood plain zone

Flood plain zone

Where the river flow slows right down, deposition is predominant. The valley is now wide, unconfined and the slope is gentle toward the sea.

The meandering channel is able to migrate laterally, followed by a sequence of point bars or scroll bars. As the meanders become giant loops, the river seeks the shortest route across the narrow neck. The abandoned loop is then separated from the main channel by deposition and an oxbow lake forms.

Water flooding the river banks dumps coarse materials, and ridges, or levees, are formed alongside the channel, which provide only a temporary barrier to further flooding. Once breached, a crevasse in the levee allows a surge of water to send a fan of material across the flood plain as a crevasse splay. The resulting marshy area is called a back swamp. A yazoo tributary travels along parallel to the river, sometimes for many miles, until it breaks through a weakness in the levee to join the mainstream.

A wedge of sediment pushes out into the sea as the river's life ends in a multitude of distributaries threading across its delta.

Rivers at work/2

Over its whole course, the action of a river varies between erosion, transport and deposition. Sometimes it cuts into its bed, sometimes it fills its channel; it depends to a large extent on the speed of the flow of water. Although a river is able to erode throughout its entire length, it has its greatest effects where the water flows fastest. Typically, this occurs in the upland or mountainous regions where the slope is steepest and water inputs are highest.

Three erosion mechanisms are at work: rivers dissolve, carry away and abrade. Susceptible rocks like limestone can be corroded and dissolve in water, and the products are removed in solution. The force of the water can sweep away loose sediment by friction at the base of the flow. Armed with sediment, the river can then scrape away at the material of its channel by abrasion. During this process the sediment itself is worn down by the constant collision of the particles.

Material too heavy to be carried in the flow is transported as bedload, which rolls, slides and bounces along the channel floor. Some large stones and boulders may only move during high velocity floods, but the finer silts and clays are held suspended in turbulent flows or eddies.

Yellowstone Falls of Wyoming, U.S., embrace two mighty waterfalls. The upper falls are 109 ft (33 m) high, while the lower falls descend an impressive 308 ft (94 m), in a narrow cascade of water that plunges into a deep and narrow chasm. This "Grand Canyon" of Yellowstone National Park forms a steep-walled valley with only enough room for the turbulent river waters.

The sediment-carrying power of a river is exemplified by the Yellow River, or Huang He, China's second longest.

From its source in the Kunlun Mountains in Tsinhai province, the river flows some 2,900 miles (4,670 km) to the Gulf of Chihli in the Yellow Sea. Scouring a succession of basins filled with loess — a deposit of wind-blown coarse silt — the river picks up vast quantities of sediment on its way.

The river's broad channel becomes so heavily laden with its suspended load of silt that the waters take on an opaque, terracotta hue.

Frequent flooding has deposited the silt along a wide flood plain and delta, creating some of the most fertile land in China, covering 35 million acres (14 million hectares).

As the river continues, more tributaries join, bringing more material with them. Once loaded with sediment the river transports it in two ways: invisibly, with the dissolved sediment dispersed throughout the flow; or suspended as individual particles. When the flow slows, the heavier particles are likely to be deposited and settle. This may be in the quiet waters of a river bend, or where a channel widens to slow flow or where an obstruction in the channel causes speed to slacken slightly.

Deposition is greatest in the lower, slower reaches of the river where the slope of the land is less, especially where a river meets the sea. As might be expected, the coarser material is first to leave the clutches of the river, with ever smaller sized particles following further downstream. This explains why rocks are found in the bed of a ravine whereas fine silt lines a flood plain or estuary.

Deltas: where rivers end

A river ends its existence with an act of creation. As it enters the ocean, the material "stolen" by the river along its route becomes the foundation for building new land, a delta. As clays, silts and sands are dropped on to the ocean floor, a delta may form.

Not all rivers produce deltas; deltas do not form where deep water or submarine canyons lie immediately offshore, as at the mouths of the Amazon and Congo rivers. If the currents along the coast are strong, silts and sands are swiftly removed from the river mouth; if cold, muddy water flows into a warm sea, the dense water sinks and sediments are spread widely.

Delta shapes vary greatly, but most can be explained by the delicate interplay between the rate at which sediment is brought down by the river and the strength of the waves, currents and tides. If the river water is much lighter than the sea water with which it merges, as when a river enters cold ocean, some sediments may be supported and carried long distances from the outlet. This encourages the construction of elongated or bird's-foot deltas, as typified by the Mississippi.

Where offshore currents are powerful, as at the mouth of the Colorado River, the sediment is quickly distributed by currents, so that deep channels form. Another pattern emerges where wave action is strong, for example in the Rhône Delta in the south of France. As sediment is deposited at the river mouth it is swept sideways and creates spits, bars and sand ridges.

Despite problems of subsidence, poor drainage and flooding, many cities have developed on deltas, including Calcutta, Alexandria, New Orleans and St. Petersburg. Fertile agricultural land, a productive marine environment for fishing, as well as ease of navigation have far outweighed the inherent problems and risks.

Surprisingly, deltas can also form inland many miles from the sea. In Botswana, in southern Africa, a delta has formed where the Okavango River drains into the basin of the Kalahari Desert. Wild fluctuations in rainfall mean that the delta completely disappears for much of the year.

The Mississippi Delta began as an outfall into the Gulf of Mexico over 100 million years ago and since then has changed route about five times. From about 2500 B.C. the delta covered 186 miles (300 km) along the Louisiana shore and nearly 62 miles (100 km) inland. Today's delta was formed by the switching from side-to-side of one of the main channels. Each channel followed the path of least resistance, carving out a widening flood plain as it went. By the 1500s the river had formed a "bird's-foot" delta.

If humans had not intervened in the 19th century and maintained the channels through sophisticated flow systems, the river would have swung south again.

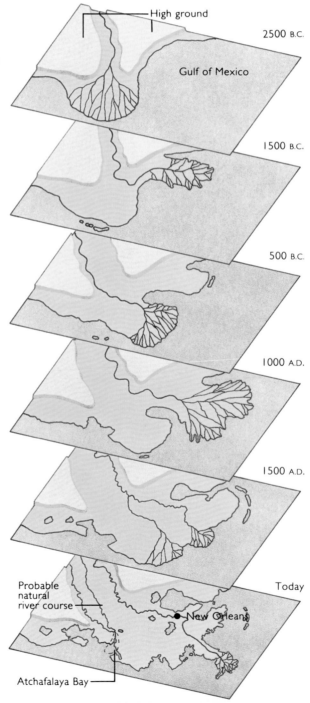

High ground

Gulf of Mexico

2500 B.C.

1500 B.C.

500 B.C.

1000 A.D.

1500 A.D.

Today

Probable natural river course

New Orleans

Atchafalaya Bay

The Nile Delta is fed by two rivers, the Blue Nile and the White Nile, which both start their lives far away in the Nubian Desert and the mountains of Ethiopia. It is down these two rivers that the sediment is carried to be deposited eventually at the mouth of the Nile in a sweep of rich alluvial mud.

Below Cairo the delta proper begins and the classic Greek letter-shape Δ is produced by two distributary rivers, the Damietta and the Rosetta. Confined between the Cairo Hills and the plateau of the Western Desert, they have steadily built outward into the Mediterranean. Here weak tides allow the waves to shape the coast into beautiful spits, complicated inlets and double bays which sweep back from the mouth.

Agriculture has always been of prime importance and many irrigation channels can be found crisscrossing the fertile alluvial mud. This inverted triangle of land holds one of the longest records of human activity on Earth. Within the delta can be found distinctive traces of the old riverbeds and sites of occupation which may be as early as the Fourth Dynasty.

The river's flow is now controlled by the Aswan High Dam, completed in 1966. Besides providing electricity, it controls the river for irrigation and water is dammed up in a lake 175 miles (280 km) long, which is starving the delta of new sediment.

Consequently, waves and currents in the Mediterranean Sea are carrying away more silt than is being replenished by the river and this is allowing erosion of the delta front.

Watery worlds

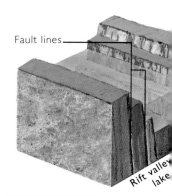

Fault lines

Rift valley lake

Areas of still water on land range from vast, serenely beautiful lakes to organically rich swamps and other wetlands. Lakes develop where water flows into and collects in a hollow or depression with a non-porous surface, or where it is trapped behind a dam. Most depressions and dams result from movements of the Earth's crust or from glacial activity. Gradual uplift of the crust can cause damming, while fault movement can create vast, deep lakes. The Rift Valley, which cuts through Africa and Asia, has led to the formation of the largest group of fault-created lakes, like the Dead Sea and Lake Nyasa.

Glaciers not only scour hollows in the ground but also deposit ridges of sediment where they melt. These ridges form natural dams behind which water collects. The Great

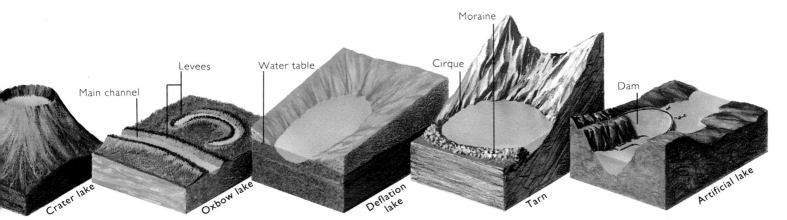

Main channel • Levees • Water table • Moraine • Cirque • Dam

Crater lake • Oxbow lake • Deflation lake • Tarn • Artificial lake

A lake is essentially a water-filled hollow. The type of lake that develops is determined by how the hollow was created.

Violent movement of the Earth's crust can result in a narrow wedge of land dropping into a steep-sided chasm, forming a long, narrow and seemingly bottomless **rift valley lake**.

A **crater lake** occupies a now extinct volcano; the collapse of a cone can leave a simple circular depression ready to be filled with water.

As meandering rivers sweep across a flood plain, the almost circular loops are cut off and abandoned for a more direct route. This crescent-shaped arm is called an **oxbow lake**.

The erosive power of desert winds can scour out a hollow. If this reaches the groundwater it can create a **deflation lake** or an oasis.

On steep mountain sides a **tarn** occupies a deep rounded basin, or cirque, formed by the scooping action of localized glaciers.

An **artificial lake** is formed when a dam is built to regulate the flow of a seasonally flooding river to ensure a constant supply of water.

Surrounded by high mountains capped with snow, Lake Louise in Alberta, Canada, is fed by meltwaters from the Rocky Mountains.

Lakes of North America and many alpine lakes in Europe were caused by glacial scouring and ridge damming as this process is called.

Most lakes are filled by rivers and streams, but some water comes from lake-floor springs and rain, sleet or snow falling on to the surface. A lake's dimensions depend on the size and shape of the hollow being filled and on the balance between inflow and outflow of water. However, levels fluctuate because of seasonal variations in rainfall and evaporation.

Depending on the balance between water inflow and outflow there are, in fact, two different types of lake – freshwater and saline, or salty, lakes. Freshwater lakes occur where there is a free flow of water through the lake; where, for instance, water collects behind an obstruction but once its level rises high enough, is able to get round or overflow the obstruction and continue onward. Freshwater lakes are vital to life on Earth, since they contain 95 percent of the surface supply of fresh water. As much as 75 percent is held in just one lake – Lake Baikal in Siberia.

Saline lakes form where the incoming waters cannot flow on to the sea. All rivers and streams carry small amounts of dissolved salts. Where there is no escape from a lake, evaporation removes water to the air leaving salts behind in an ever more concentrated solution. The Dead Sea is the world's saltiest lake, about nine times more saline than the ocean, which means that swimmers float in it like corks.

Waters pouring into a lake carry sediments as well as salts, and sedimentation can spell the death of a lake. The gradual encroachment of sediment from the shore to the centre reduces the lake to a muddy, boggy hollow and then to a rich vegetated swamp or wetland.

Watery worlds/2

In a mangrove swamp in Miami, Florida, a red mangrove sends out a web of stiltlike roots into the soft submerged mud. Mangrove swamps are rich, diverse and productive habitats where the arching roots provide stability in the tide, trap shifting sediment and make shelter for many aquatic animals.

Diverse examples of wetland are found in the flooded edge of the Panhandle in the upper delta of the Okavango River, Botswana. The white sand in the centre is all that survives tidal scouring of the low tide area.

The high zone is characterized by a dense cover of vegetation, which encourages further build up of sediment. The lower reaches of the swamp are colonized by plants which can put up with the alternating fresh and salt water of the tide cycle. A scrub type vegetation colonizes the better drained high tidal flats.

Neither truly land nor water, swamp areas and wetlands contain a uniquely rich environment. Inland wetlands form from silted lakes or from flood plains as in the reed and rush swamps of Africa and Asia. But most wetlands are found at the fringes of the seashore and in estuaries.

Waters driven by the endless ebb and flow of the tides and frequent floods inundate these areas, diffusing through a vast maze of shallow branching inlets and coves. Fed by sediment-rich rivers and tides, the quiet waters between the mean low and high tide and flood marks form expanses of mudflats. As sediments accumulate, parts of the flats are exposed. In the flats farthest from the rising waters, gently sloping surfaces provide stability for plants to colonize: marshes support a mosaic of plants, each occupying a particular niche according to moisture levels and its salt tolerance. Near channels, shifting sediment and almost permanent immersion in water proves alien to all but the most hardy plants.

The mangrove swamps, with their eerie submerged root-forests, are probably the best-known swamps. They are found only in subtropical regions, where they form the most productive and diverse ecosystems in the world along coasts and river mouths.

Underground lakes

Planet Earth consists predominantly of water. The total amount does not change, but its form changes constantly. Whether ocean or ice, cloud, snow, rain, spring, river or lake, all water is part of a single, dynamic system. It is perpetually on the move, to the next part of its unending cycle between land, sea and air.

Less than 3 percent of the water is fresh, and 75 percent of that is frozen in ice sheets. Of the balance, 95 percent is stored underground in the form of groundwater. Most groundwater originates as rain or snow. It percolates down through a network of pore spaces between soil particles, washing over hard rock, such as granite, and seeping down again when it encounters beds of sand or gravel, or porous rock that holds water, such as sandstone or chalk. Eventually it hits the surface of a vast groundwater system, below which all the pore spaces are full of water.

At the bottom of this underground lake, called an aquifer, is an impenetrable layer of hard rock or clay. Its surface – the water table – is not level but a blurred reflection of the contours of the land above. As rivers flow downhill, groundwater flows down the slope of the water table. Slowly but inexorably, a few inches a day, it makes its way to the ocean, sometimes travelling hundreds of miles from where the rain originally fell.

Not all areas have aquifers, and not all the groundwater reaches the ocean, or not by the underground route. Some seeps into rivers or lakes, and some escapes back to the surface via springs; water gushes out of Silver Springs in Florida at 6,000 gallons (23,000 litres) per second. Large amounts of groundwater are captured by deep-rooting plants, which carry it to the surface and return it to the atmosphere, as water vapour; the water table falls slightly during the day, when plants absorb water, and rises again at night, when they stop.

Aquifers accumulate their stores of groundwater over thousands of years or longer. Some of the water they contain may even date back to the earlier, wetter climate phases. Many desert regions are able to tap deep reserves that have accumulated over 40,000 years ago. In some arid regions, however, these bodies of water are unlikely to be replenished.

The underground lake is a precious resource, which must be conserved, but in many areas the groundwater level is falling by 10 feet (3 m) a year because of over-extraction through deep wells. Twenty million Americans depend for all their water on the Ogallala Aquifer, under the central United States. It covers 156,000 square miles (404,000 sq km) but is fast being depleted. Today 150,000 wells tap into the reserves, but the climate is now too dry to replenish it.

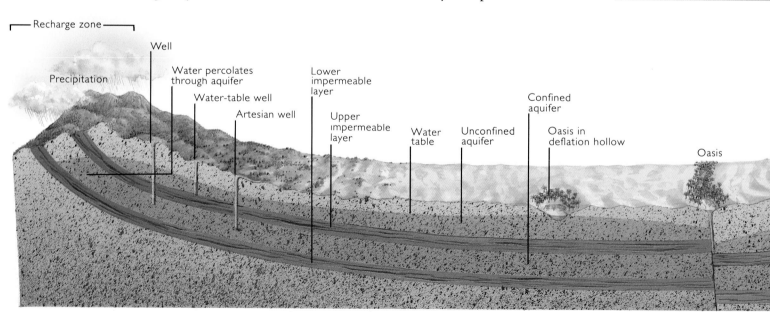

Recharge zone

Well

Water percolates through aquifer

Precipitation

Water-table well

Artesian well

Lower impermeable layer

Upper impermeable layer

Water table

Unconfined aquifer

Confined aquifer

Oasis in deflation hollow

Oasis

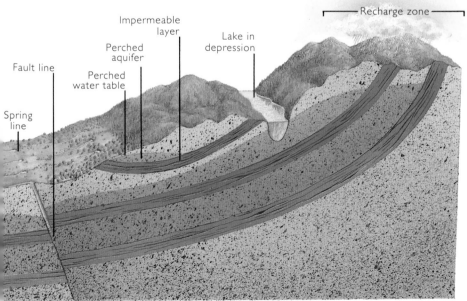

Water percolates deep underground through porous rock until it hits an impermeable layer forming an unconfined aquifer.

Water trapped under pressure between two impermeable layers forms a confined aquifer. When a well is sunk here, water shoots to the surface in an artesian well. An artesian spring arises above a fault.

An impermeable layer above the main water table may trap water in a perched aquifer. If a valley or depression reaches below the water table, a lake forms. In desert areas, oases may spring up here.

The Souf Oasis in the Algerian Sahara (above). Concentric walls of closely planted vegetation hold back the ever advancing sands from swamping the date palms growing in the center of the oasis.

Diagram labels:
- Fault line
- Spring line
- Perched water table
- Perched aquifer
- Impermeable layer
- Lake in depression
- Recharge zone

Water: the master sculptor

Nature takes her time. To chisel rock, she uses water, drop by drop, skimming off mere molecules with each stroke. Given millions of years, this patient nibbling carves out spectacular monuments such as the Carlsbad Caverns in New Mexico, where no end has yet been found, or the hanging hills of Guilin in southwest China, jutting like giant dragon's teeth hundreds of feet above a flat plain.

First came the rock – limestone, which is formed with a similar lack of haste. This sedimentary rock consists of the fossilized remains of creatures that lived in shallow, warm seas during the time of the dinosaurs, a hundred million years ago. Plankton skeletons, sea shells, coral and detritus gradually built up into thick beds on the sea floor, which eventually fossilized into limestone. Since then, the restless shifting of the Earth's crust has rearranged the continents, thrusting up most of today's great mountain ranges and leaving former seabeds on dry land.

Limestone is a porous rock: its fine-grained internal structure holds water, like a hard sponge. Most of the world's groundwater is stored in ancient, buried seabeds of limestone. The horizontal beds are built like a brick wall, with layers and vertical joints giving it

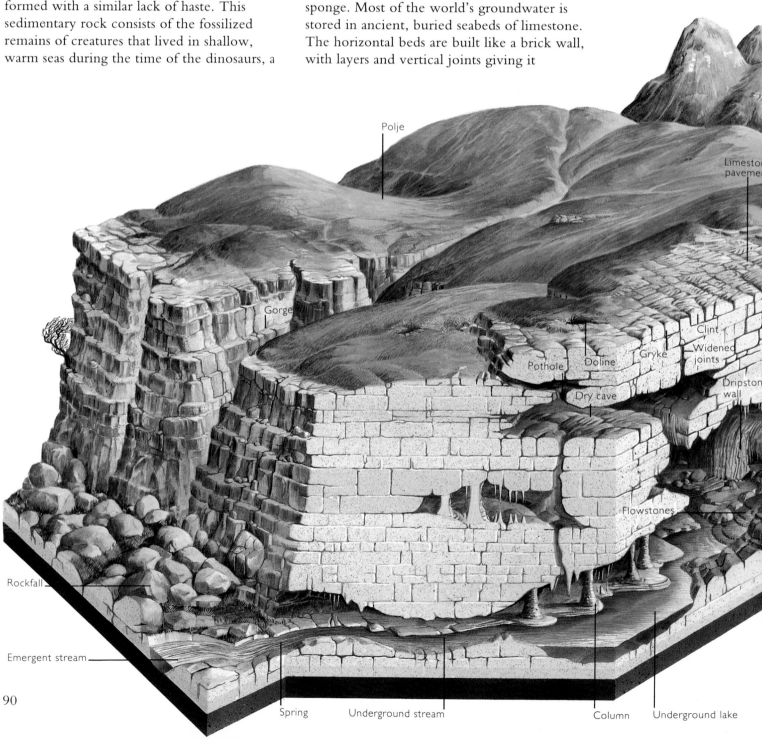

Polje

Limestone pavement

Gorge

Clint

Pothole

Doline

Gryke

Widened joints

Dripstone wall

Dry cave

Flowstones

Rockfall

Emergent stream

Spring

Underground stream

Column

Underground lake

structural strength. The joints and parallel cracks caused by faults form a network that water seeps into.

As rain falls, it picks up carbon dioxide gas from the air, so forming weak carbonic acid. It gathers more carbon dioxide as it filters through the soil, as well as organic acids. When the water encounters limestone, the acids convert some of the lime into calcium carbonate, which dissolves in the water and neutralizes the acid. The next raindrop eats away a little more lime before it, too, loses its corrosive power.

Water seeps fastest and farthest through the joints and cracks and along the bed layers, cutting the stone into pavements of even-sized slabs, or clints. The joints, or grykes, between them are gradually eroded by the water's action.

Water then concentrates at junctions in the rock where major joints converge, slowly eating out a steep-sided hole in the ground, called a doline. Dolines deepen, spread and join, until eventually there is more doline than land. The limestone landscape that is found in parts of Jamaica looks like an upturned egg-tray: the tops of the conical hills are all that remain of the original land surface. As long as there is rain, the land surface keeps sinking.

Residual limestone hills

Sinkhole

Chimney

Galleries

Stalactite

Stalagmite

Impermeable layer

The often barren surface of a karst landscape (named after a region in Slovenia where these features occur) often gives no clue to the spectacular network of underground channels or caverns below.

Water seeps down from above and gradually sculpts the limestone into a vast number of potholes, underground corridors, caves and caverns, sometimes with waterfalls, lakes and rivers running in them.

Once the water has found its natural level, or hit a lower impermeable rock layer, it may form an emergent stream or underground lake. When the stream has dried up or altered course, dry galleries remain, which may come to be adorned with stalactites and stalagmites.

A stalactite grows down from the cave ceiling in Carlsbad Caverns, New Mexico. Dripping water deposits a calcium carbonate solution which increases in size to form a stalactite. Deposits also build up where the drops splash on the floor, growing into a stalagmite. The two "growths" often meet, merging into columns.

Water: the master sculptor/2

The cones become eroded into tower karst, tall spires of limestone like those in Canada's MacKenzie Mountains.

Streams flowing over karst can vanish into sinkholes. Beneath the surface, the water continues its work. Rivulets running through the joints join up to form single streams, cutting broader channels. After a long time and much rain, the streams cut out a limestone cave. For example, it took about 60 million years to carve Carlsbad's labyrinth of interconnecting corridors and caverns, which drop to a maximum depth of 1,100 feet (335 m).

The world's longest cave system, Mammoth Cave in Kentucky, has 329 miles (530 km) of passages, on five levels. The largest known underground cavern is Sarawak Chamber in Sarawak, Borneo. At 2,297 feet (700 m) long by 1,312 feet (400 m) wide, eight jumbo jets could park nose to tail across its middle. Because the limestone beds out of which it is carved are thick and unscarred by joints, it is totally unsupported other than at its sides.

The water in limestone chambers is at its most corrosive when it reaches the surface of the groundwater system, the water table. If the water table falls, the caves drain and the rainwater cuts new channels at the lower level. Rainwater usually keeps on seeping into empty caves, having percolated through the soil where carbon dioxide concentrations are 100 times higher than in the open air.

Exposed to the air in the cave, the dissolved carbon dioxide in the water evaporates, and the water loses the calcium carbonate it has absorbed from the limestone, leaving it on the cave ceiling or on the floor. Where the mineral load is left on the ceiling, stalactites hang down. Stalagmites grow up from the floor when the water drip has been more rapid, dropping to the ground before depositing its calcite. Dripwater oozes over the cave floor glazing it; chemical impurities in the water add orange and brown stains, especially where water trickles down a wall, and leave a translucent curtain of stone sometimes just one crystal thick.

Limestone cave systems are still being discovered worldwide and many surprises await the explorer of these amazing worlds.

Throughout China's long history, few major artists have not painted the famous hills of Guilin, in the southern province of Guangxi.

The epitome of the classical Chinese landscape, the steep pinnacles jut out of the flat flood plain of the River Xi, rising to 330 feet (100 m). Stunted trees cling to craggy outcrops, and vines hang like garlands down the cliffs. The hills accompany the river for 30 miles (48 km). Their tops are the remnants of the former ground level – the rest of the land has been eroded by water.

The Guilin Hills are the world's finest example of tower karst, thought to be the final stage of an eroded limestone landscape. Guilin has a high rainfall, and over time the rain ate away at joints in the limestone, forming a grid of deep, narrow gorges. The water then worked at the cubes of rock between the gorges, leaving the thin spires of today standing over a wide, flat plain, the weathered ribs of a vanished landscape.

Sweltering sun and swirling sand

The popular picture of a desert is of a desiccated inferno. But nature is much more diverse than this and deserts range from the blazing tracts of the African Sahara to the frozen wastes of Antarctica. Deserts may be hot during the day and cold at night, such as the Kalahari in Africa, or they may be permanently cold, such as the Taklamakan Desert in China.

Deserts can be found on every continent; their only common feature is that they have little rainfall – less than 10 inches (250 mm) a year. Four factors contribute to low rainfall and hence the formation of deserts: high pressure areas; cold, upwelling currents; mountain rain shadows; and a location in the middle of a continent.

Of these, the most significant is high pressure. This comes about when the air is heated by the Sun at the equator. Warm, moist air rises and flows outward toward the tropics. As it rises, it cools, condenses and sheds moisture before sinking back to Earth at

between about 15 and 30 degrees latitude in each hemisphere. The descending air is compressed and warmed by the air pressure above it, which increases its capacity to retain moisture. As a result, little precipitation falls, there is little cloud cover and water on the ground evaporates quickly because the Sun beats down without interruption.

Cold water currents welling up from the ocean depths along the western coasts of subtropical landmasses create cold surface water. This chills the air and reduces the amount of water it holds and stops rain reaching land. Deserts created this way are the Atacama in Chile and the Namib in southwestern Africa.

The third cause of dry air is the rain shadow effect. When moist air moving inland from the sea is forced upward by mountain peaks, the air cools and its water vapour falls as precipitation on the windward slope. Only dry air continues inland, warming as it descends and forming a desert in the rain shadow.

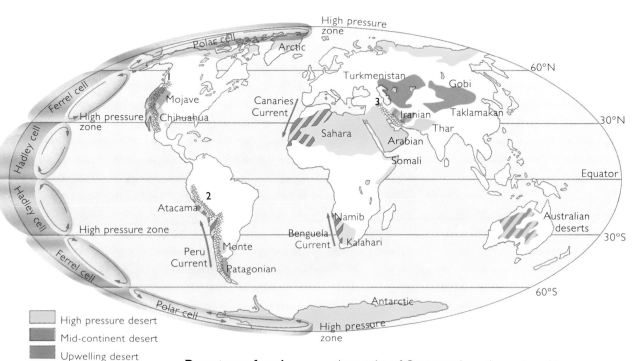

Deserts are found wherever there is little precipitation in the form of rain, sleet or snow.

Many of the Earth's deserts are concentrated under two zones of high pressure air which bracket

the tropics of Cancer and Capricorn. These high pressure deserts are found on all continents.

Other deserts are created where air has had time to lose all its moisture before it reaches

the centre of large continents. This type of desert is dominant in Eurasia and Australia; in the latter, high pressure is also a contributing factor.

Cold currents welling up along the western coasts

of continents are responsible for creating some of the deserts in South America and Africa.

Where air has had its moisture extracted by being forced up over high mountain ranges, deserts

Legend:
- High pressure desert
- Mid-continent desert
- Upwelling desert
- Rain shadow desert
- Mountains forming rain shadow
- Cold upwelling
- 1 Sierra Nevada
- 2 Andes
- 3 Zagros Mountains

form on the leeward side in the rain shadow of the mountains. The Mojave Desert in the rain shadow of the Sierra Nevada contains Death Valley, the hottest place in North America. Temperatures can reach a searing 131°F (55°C) and rainfall is just 1 inch (25 mm) a year.

Antarctica is the driest place in the world – even drier than the Sahara Desert. There are three main Dry Valleys, gouged out by glaciers long ago. Today, as for the last two million years, only a scant dusting of snow falls. The result is a bitterly bleak, cold and arid landscape, where gusting winds prevail.

An Egyptian oasis in the Western Desert, part of the Sahara, struggles to resist the relentless march of the dunes. Many villagers have left to seek a new life elsewhere.

Sweltering sun and swirling sand/2

A good example is the Mojave Desert in the shadow of the Sierra Nevada, California.

The final factor in the genesis of deserts comes about because by the time airflows reach the centre of a large landmass they have shed all their moisture. Examples include the Gobi in Mongolia, the Turkestan in Central Asia and the Australian deserts.

As with so many landscapes, it is the power of wind and water that sculpts the desert. Water – even in deserts – is an effective agent of erosion. During storms, sparse vegetation, rocks and compact, crusted soils encourage sheets of water to flow over the surface and this sheetflow collects to form a network of rills and gullies. In large channels, or wadis, destructive torrents of water can remove vast quantities of debris. Where mountains meet plains, debris off high ground is dumped as cone-shaped alluvial fans or as a continuous skirt of coalesced fans, called a bajada.

Wind erosion is divided into abrasion, where sand grains hit a surface and wear it down, and deflation, where wind blows away small-grained material. Abrading particles have the power to etch, polish and groove rock and pebble surfaces. These wind-created forms, called ventifacts, include the smallest pebbles as well as large streamlined rockforms called yardangs, which range from features only yards high to ridges of 330 ft (100 m). Iran's Lut Desert is famous for yardangs.

Desert pavements, or regs, form when mixed deposits are attacked by wind, which slowly removes the small particles to leave a layer, or lag, of pebbles and stones. This coarse sediment is concentrated to form a pavement of closely fitting pebbles. In the Sahara, pavements stretch for up to 1,000 miles (1,600 km).

The wind is also responsible for the most recognizable feature of sand deserts – dunes. Sites where wind-blown sand is deposited and accumulates are called ergs. Dunes can form when streams of moving sand encounter an

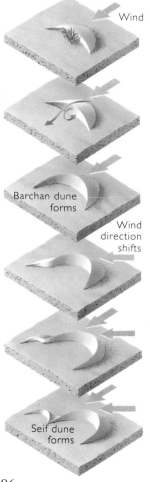

Barchan dunes can only form in deserts in which winds blow constantly from one direction. They are concave on the side sheltered from the wind.

By contrast, seif dunes form when there are regular winds from two points of the compass. One wind blows across the dune axis to pile sand up; the other wind blows along its axis to blow sand out of hollows between the dunes.

These seif dunes in Scarmann's Lagoon, Baja California, Mexico, show their scimitar shape particularly well when seen from above.

Bryce Canyon in Utah is a vast amphitheatre of multi-coloured rocks given their hue by oxidation of metals such as iron and manganese. Formed from huge dunes piled up by the wind 200 million years ago, the sandstone layers have been eroded into dramatic forms.

obstruction, such as a boulder, vegetation or irregular ground; but they can also accumulate and grow on a variety of surfaces.

The type of dune depends on the supply of sand, the direction of the wind, the presence of any vegetation and the shape of the terrain. There is a multitude of shapes and sizes including barchan, or crescent-shaped, dunes; star dunes forming sand peaks; and dome dunes. There are also linear, or seif, dunes which look like a sword edge with steep-sided ridges following the wind.

Although some dunes are fixed, anchored by objects which disrupt the path of the sand and wind, others move quickly. Barchan dunes are the most mobile, moving up to 100 ft (30 m) a year, while linear dunes edge along by as little as half an inch (1 cm) a year.

Ice and fire

Continental ice sheets have shrouded large parts of the planet for a total of 50 million years or more. In terms of the Earth's long history, however, ice ages are rare events. Rocks in the Gowganda Series in Canada were deposited more than 2,000 million years ago during what is thought to have been the first ice age. The next one occurred 1,300 million years later.

More recently, however, they have been far less rare. Earth has been almost constantly icebound since the start of the Pleistocene epoch two million years ago. Glacial ice sheets 2 miles (3 km) thick have reached as far south as New York City five times. There have been 10 major glacial periods and up to 40 shorter periods of glaciation. The ice generally stayed for 80,000–100,000 years, giving way to warmer interglacial periods lasting 10,000–20,000 years.

Humans were born during the last ice age, and only started to become civilized when it ended, 10,000 years ago. The current period is seen as an interglacial period, which will eventually give way to a new ice age.

A vivid picture of the last ice age can be obtained by piecing together the record found in the two remaining ice sheets in Antarctica and Greenland, in rocks, fossilized pollen and microfossils in ancient levels of the seabed.

It arrived suddenly, with dramatic effects. The climate changed the world over, fluctuating wildly, but with overall cooling. Winds grew stronger, rain and snow increased, clouds built up to reflect the Sun's heat away from the Earth's surface. World temperatures fell 5–14°F (3–8°C); and ocean surface temperatures dropped 4–11°F (2–6°C), further cooling the air. The altitude at which glaciers can form was reduced by 5,000 feet (1,500 m). The polar ice caps spread and ice sheets formed in the continental highlands, finally covering most of the temperate regions. More than a third of the land lay under a blanket of ice 8,000 feet (2,400 m) thick.

As the ice sheets grew, ocean levels dropped, by as much as 200–400 feet (60–120 m), as their waters were bound up in ice. Whole coastlines moved out 100 miles (160 km) or more: Britain was joined to Europe, and Alaska to Siberia.

The Earth has been in the grip of an ice age for most of the last two million years.

The graph shows temperature changes over the past 850,000 years. During the ice ages, each of which lasted up to 100,000 years, thick ice sheets advanced from their centres at the poles to cover as much as a third of the world's surface.

The interglacial periods were shorter, lasting 10,000–20,000 years, with warmer temperatures similar to today's mean of 59°F (15°C).

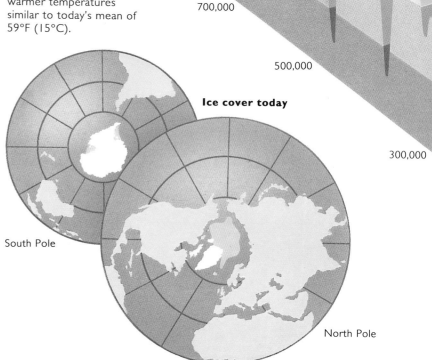

850,000

700,000

500,000

Ice cover today

300,000

South Pole

North Pole

Milutin Milankovitch, a Yugoslavian geophysicist, suggested in 1924 that the sequence of ice ages was the result of slow changes in the Earth's orbit about the Sun.

He calculated that the combined effect of what has since been termed the "wobble, roll and stretch hypothesis" would cause temperature changes corresponding to the glacial advances and retreats that occurred in the Pleistocene epoch.

Orbital eccentricity

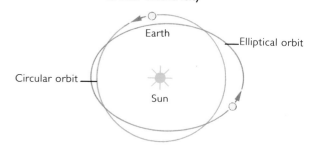

Earth

Elliptical orbit

Circular orbit

Sun

Over a 97,000-year cycle, the Earth's orbit changes from being a circle to a slightly elliptical shape – the "stretch factor". This eccentricity of orbit causes seasonal variations in the amount of heat that is received from the Sun.

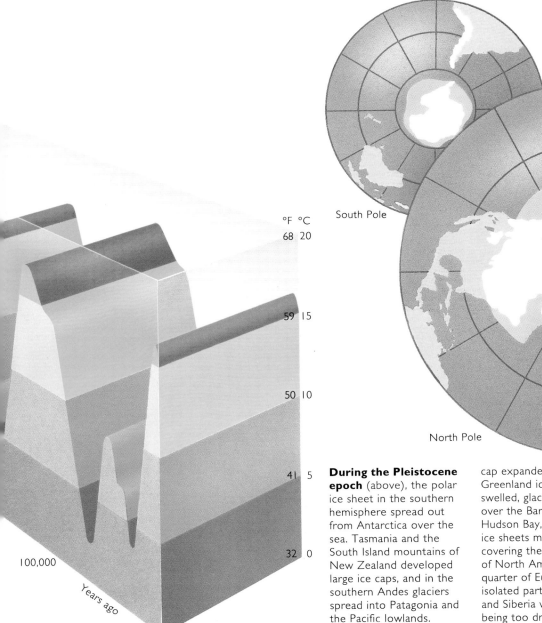

Ice cover during last ice age

South Pole

North Pole

°F °C
68 20
59 15
50 10
41 5
32 0

100,000

Years ago

Today

During the Pleistocene epoch (above), the polar ice sheet in the southern hemisphere spread out from Antarctica over the sea. Tasmania and the South Island mountains of New Zealand developed large ice caps, and in the southern Andes glaciers spread into Patagonia and the Pacific lowlands.

In the northern hemisphere, the polar ice cap expanded, the Greenland ice sheet swelled, glaciers spread over the Barents Sea and Hudson Bay, and highland ice sheets moved south, covering the northern half of North America and a quarter of Eurasia. Only isolated parts of Alaska and Siberia were ice-free, being too dry for glaciers.

Today, two ice sheets remain (far left). Nine-tenths of the world's glacial ice is in the Antarctic ice sheet, which covers an area bigger than the U.S. and is up to 13,000 feet (4,000 m) thick. The Greenland ice sheet is much smaller.

Many of the world's mountain ranges still have glaciers in the higher peaks and valleys. Glaciers even survive in Africa; only Australia has none.

Axial tilt

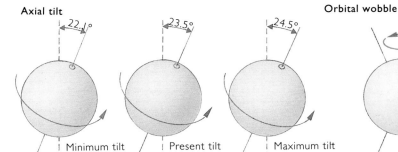

Minimum tilt Present tilt Maximum tilt

Orbital wobble

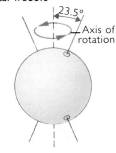

Axis of rotation

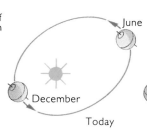

June

December

Today

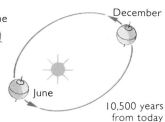

December

June

10,500 years from today

Every 41,000 years the tilt, or roll, of the Earth's axis moves between 24.5° and 22.1°. The greater the tilt, the more marked the seasons. The polar regions receive less sunlight when the tilt is decreased and more sunlight when it is increased.

The Earth wobbles in space like a slowly spinning top. This means that the time of year at which the Earth is nearest to the Sun (perihelion) varies. At present the northern hemisphere winter is in perihelion, while the summer occurs at the farthest point in the orbit. Since the cycle takes 21,000 years to complete, in 10,500 years the position will be reversed.

Ice and fire/2

The winds carried away the exposed seabeds, filling the air with 10 times as much dust as there is today.

In the tropics, beyond the walls of ice, temperatures were lower and it was much drier than today; the subtropics were wetter, with large lakes in north Africa. In North America, the ice scoured out the Great Lakes.

The interlocking plates of the Earth's crust sank under the tremendous weight of the ice sheets. When the ice receded, the crust started to spring back to its former position – a gradual process that is not yet complete. Areas such as Canada and Scandinavia, which were below the centres of ice sheets, are still rising. Canada is gaining land every year and could rise a further 1,300 feet (400 m).

What is not at all clear to scientists is how ice ages are triggered. One theory is that the Pleistocene ice age was caused by continental drift. Africa, South America, Australia, India and Antarctica were once part of the southern landmass Gondwana, while Eurasia and North America, combined in Laurasia, were near the equator. Gondwana broke up, and the new continents drifted north, while the northern continents closed in around the Arctic Ocean. This effectively cut off the Arctic seas from the warm ocean currents flowing from the equator, and the climate cooled.

Continental drift is slow: it helped pave the way for an ice age, but it does not explain the sequence of drastic swings between glacial and interglacial periods. In 1924, a Yugoslavian geophysicist, Milutin Milankovitch, proposed a different theory, based on slow changes in the Earth's orbit around the Sun. He calculated that these long cycles would cause temperature changes which fitted the glacial cycles of the Pleistocene ice age.

The theory is now widely accepted. Recent studies of fossil remains in core samples taken from ocean beds indicate a sequence of temperature changes which closely match the Earth-Sun cycles. Milankovitch's solar ice-age horoscope predicts that the cycles will continue – the ice sheets will return. Estimates of when that will happen vary between 23,000 years from now and much sooner.

The Vignemale Glacier
in the Western Pyrenees
National Park, France, is
one of the few remaining
glaciers found in this
mountain range.

During the recurrent ice
ages of the Pleistocene
epoch, the altitude at
which glaciers form fell by
as much as 5,000 feet
(1,500 m) and, like many
other mountain glaciers,
the Vignemale extended
far down the valley. In
retreat during the present
warmer period, the glacier
is now confined to the
steep slopes below the
frost-shattered peaks, at
an altitude of nearly
11,000 feet (3,300 m).

The relative size of the
people in the photograph,
crossing the glacier in the
centre, gives an idea of its
extent.

Rivers of ice

Glaciers are among nature's most powerful tools. They carve spectacular landscapes and can grind mountains to dust – the proverbial mills of God. Yet they are made out of snow.

Glaciers form above the snow line, from layers of snow that survive the summer melt. As the snow builds up, the snowflakes go through a sequence of compaction, melting and refreezing that transforms them into ice. Over 4 million cubic miles (17 million cu km) of glacial ice covers about 11 percent of the land surface. Most of it is in the enormous ice sheets of Antarctica and Greenland, and the rest in thousands of mountain glaciers.

There are more than 100,000 mountain glaciers in the world, found on all continents except Australia. They originate where snow accumulates at the head of a high valley, flowing down the valley like a river. The sheer weight of a glacier deforms the ice crystals, which slip against each other, allowing the ice to flow. A few glaciers surge at up to 60 yards (55 m) a day, but most creep just a few feet a day. In Antarctica, for example, glaciers may move less than 3 feet (1 m) in a year.

At the base of the glacier, where ice meets bedrock, the immense pressure creates a thin film of meltwater, helping the glacier to slide over the rock. Where the land is uneven, the glacier either cracks, forming deep crevasses in the surface, or buckles to form pressure ridges.

As long as there is more snow accumulating at the source than meltwater flowing out at the lower end, or snout, the glacier advances

Most glaciers begin life in cirques, ice-eroded basins in high mountain slopes. As snow builds up, ice spills over the cirque into the valley, forming a glacier, which may be joined by smaller glaciers emerging from tributary valleys.

In the upper reaches, frost-shattered rock fragments fall on to the edge of the glacier. Where two glaciers meet and join, these lateral moraines converge into a medial moraine running right down the middle of the glacier.

When the glacier hits a steeper part of the valley, internal stresses split the surface into a wedge-shaped crevasse, often 50 feet (15 m) wide and 100 feet (30 m) deep. In winter they can be covered by snow, invisible to the unwary traveller.

At the snout the glacier melts and dumps its burden of rock. Below the snout, ridges of terminal moraine document the glacier's retreat. Running almost perpendicular to this moraine are long, narrow eskers – coarse gravel ridges – formed by subglacial streams.

Meltwater streams form complex, braided channels in the outwash plain, depositing their sediment in well-sorted bands. The glacier's moraine, however, is unsorted: rocks of all sizes fall in a jumble as the ice melts.

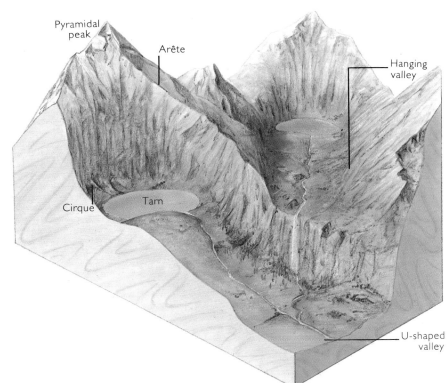

downhill. With more meltwater than snow, the snout shrinks back uphill – although ice continues to flow down through the glacier. Only when snow stops accumulating at the source does the ice stop flowing, and when there is a loss of snow at the source, the glacier decays. Many glaciers are only 1–2 miles (2–3 km) from source to snout, but the Hubbard Glacier in the Yukon is a massive 68 miles (109 km) long.

Glaciers are loaded with rock debris, or moraine, ranging from large boulders to dust. Some gathers in the upper reaches, dropping on to the glacier from the steep valley walls, but most comes from the bedrock at the base. The ice freezes on to the underlying rock and plucks out chunks of it as the glacier moves on.

When a big glacier decays, the receding ice exposes a landscape transformed. Little more than the frame of the original mountain is left. At the head of the valley, the slopes are hollowed out into armchair-shaped depressions called cirques, often filled with lakes, or tarns, and divided by jagged ridges, or arêtes.

The main valley is now a deep trough, with tributary valleys shorn off above to form hanging valleys. Often waterfalls cascade over the drop to the main valley floor. The truncated spurs above the hanging valleys are all that remain of the former valley sides.

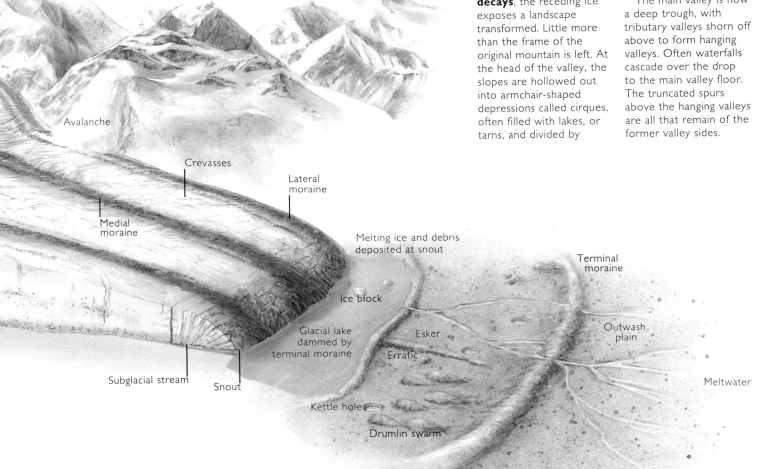

103

Rivers of ice/2

The glacier grinds these quarried fragments together, pulverizing them into rock flour, which polishes the bedrock, while larger fragments gouge out long scratch marks.

The glacier eats away at the valley, cutting a deep, steep-walled, U-shaped trough. It undercuts tributary valleys, leaving them hanging high above the main valley floor when the ice recedes. California's Yosemite Valley epitomizes the superb scenery glaciers can leave behind them.

The debris is dumped at the glacier's snout as the ice melts, leaving a ridge across the valley, known as terminal moraine. Lakes often form behind these rock walls, which can be hundreds of feet high. Retreating glaciers leave a series of ridges in their wake.

Glaciers release vast quantities of sediment-

laden meltwater, especially in summer, with powerful streams gushing out from under the snout. Meltwater streams flow at high pressure under the ice – they can even flow uphill, and carry large boulders with ease. The streams are themselves a powerful erosional force.

Streams of surface meltwater charged with sediment vanish into the body of the glacier via sinkholes, joining with subglacial meltwater to cut an underglacier network of tunnels flowing toward the snout. When they emerge from under the ice, the streams drop their burden of rock in an outwash spreading down-valley: first the larger rocks, then gravel, then sand, and finally rock flour.

They also deposit large chunks of ice, covering them with rock sediment. When the ice melts, "kettle holes" are left behind.

Lake Wanaka, a roughly triangular lake, is fed by the meltwater of surrounding rivers and glaciers. Situated at the southern end of New Zealand's Southern Alps, it is 30 miles (48 km) long and more than 6 miles (10 km) at its widest.

A swarm of drumlins at Loch Skeen in Scotland. These teardrop hills are all the same shape and they all point the same way. Drumlins (from the Gaelic word for mound) were probably shaped by two glaciers. The first, overloaded with rock, came to a halt, and the second sheared over it, shaping and smoothing the rock deposits.

Although most kettles do not exceed 1½ miles (2 km) in diameter, some in Minnesota are more than 6 miles (10 km) wide. In many places water eventually fills the depression, which is transformed into a pond or lake.

The rarest type of glacier is the piedmont glacier, which escapes from its valley to flow out over the plains. The best example is the Malaspina Glacier in Alaska, which is formed by the confluence of a number of mountain glaciers. It covers an area of 1,600 square miles (4,200 sq km) and is more than 1,000 feet (300 m) thick.

Frost in action

In the lands that rim the ice of the far north and south, the extreme winter cold combined with the summer thaw have profound effects on both the landscape and the people living there.

A major feature of these periglacial regions is permafrost. As its name implies, this is a region of permanently frozen ground extending well below the surface. In parts of Siberia and northern Canada, the permafrost can go as deep as 1,650 feet (500 m) and at its coldest is 5°F (-15°C). At the top, however, is what is called the active layer, which melts in the brief summer and then freezes with winter's return.

In permafrost regions this seemingly insignificant process – the alternate freezing and melting of water – can have dramatic effects, including the breaking apart of even the toughest rock. It is all due to one simple fact: water expands when it freezes.

Freezing and thawing of the active layer is responsible for several landscape features including seemingly regular patterns of stones on the surface and low hills, called pingos. But perhaps the most dramatic phenomenon is the frost shattering which takes place when exposed rocks become inundated with water. When this freezes the pressure exerted by an expanding wedge of ice in a crack levers the rock asunder to produce loose fragments.

Frost shattering generates rock debris which accumulates at the foot of steep slopes. These slopes are rarely stable since frost action continues to work on the scree, or talus, breaking down the individual fragments themselves. These are then moved downslope by the levering action of ice formed in the rock spaces and by snow avalanches.

In a permafrost area people must take special precautions. In fact houses are built on stilts since melting of permafrost can cause foundations to sink or crack. Oil pipelines cannot be buried in the ground as this would melt the permafrost and the pipes would sag and break. These are fragile environments and damage to the surface layers caused by mining, drilling, or driving over the surface can take a long time to heal.

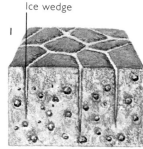
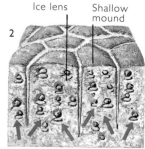
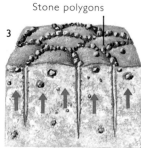

Ice wedge — Ice lens — Shallow mound — Stone polygons

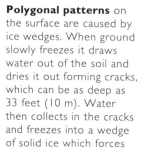

Polygonal patterns on the surface are caused by ice wedges. When ground slowly freezes it draws water out of the soil and dries it out forming cracks, which can be as deep as 33 feet (10 m). Water then collects in the cracks and freezes into a wedge of solid ice which forces

the soil apart as it expands, creating the boundary of a polygon (1).

As the ice wedge grows it compresses the surrounding sediment which bulges in response to form a shallow mound. A tiny ice lens develops under stones in the soil, pushing them upward, in a

process known as frost heave (2).

When the stones finally reach the surface, they roll down to the base of the mound, collecting in the cracks to form boundaries of polygons (3).

One of the more startling sights of Canada's Northwest Territories are the highly visible patterns at the surface made by stone polygons (far left). Here these ringlike arrangements interconnect to form a vast fishing net formation.

Conical mounds called pingos – the Innuit word for hill – are often pushed up by ice beneath the surface in areas where permafrost rules. Most pingos are found in or next to a lake basin; they are large roughly round to elliptical mounds with cores of ice. They range in size from 16 to 230 feet (5 to 70 m) high and from 100 to 2,000 feet (30 to 600 m) across.

This depression surrounded by a rampart (left) is a collapsed pingo on the Tuktoyaktuk peninsula in Canada.

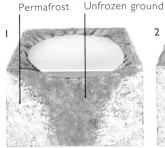

Permafrost   Unfrozen ground

1

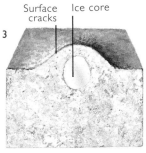

Permafrost forces water upward

2

Frozen lake

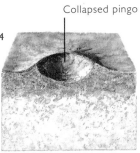

Surface cracks   Ice core

3

Collapsed pingo

4

Pingos are common features in permafrost regions. A pingo forms when a shallow broad lake insulates the ground beneath to create an unfrozen zone (1). When the lake is frozen solid permafrost encroaches from the base, top and bottom, isolating the

unfrozen zone (2).

The water in this area then freezes and the immense pressure exerted by the ice as it expands causes the overlying sediment to dome up (3). The roughly round dome-shaped hill, or pingo, has an ice core, and its summit is often marked by

a number of large cracks.

If the cracks are deep and wide enough the ice core is eventually exposed. It may then melt and the ground above caves in to make a craterlike collapsed pingo (4).

The answer lies in the soil

Geologically speaking, soil occupies an almost infinitesimally thin layer between the underlying bedrock and the ground surface. Despite this, soil is all-important to us, for in it grow the plants on which all land creatures depend, whether directly or indirectly, for food.

Soil is not all the same; it varies according to local geological and climatic conditions. The prerequisite for the formation of soil is a surface on which loose debris can collect. Where slopes are so steep that no material can gather, no soil forms. Where material can collect on steep slopes, soil tends to be thin and poor. It is only on almost level terrain that deep, rich soils can develop.

The raw material of soil is small fragments of rock called regolith. These small fragments are the end result of physical or chemical weathering processes such as frost shattering and abrasion, and can either be rock debris that comes directly from the bedrock below or material deposited by rivers, glaciers or wind. The regolith supplies the soil with mineral particles, the size of which determines the texture of the soil. Particle sizes range from coarse stones and gravels, through sand, to fine silts and clays.

Generally, the larger the average particle size the more readily a soil both absorbs and drains water. Thus sandy soils – with moderately sized particles – drain freely, whereas clay soils – with minute particles – are poorly drained. But mineral particles alone – regardless of size – do not necessarily constitute a fertile soil. Soil is an intimate mixture of mineral particles and organic matter; it also contains water and air, both vital to the organisms growing in the soil.

Soil is arranged in layers called horizons, each of which is defined according to the size and type of the mineral particles and the organic content. There are four main layers, the topmost one being the O horizon, made up entirely of organic matter such as leaf litter. This can only form where there is a good cover of vegetation.

Below this is the A horizon, or topsoil. Here organic matter has decomposed to form humus and has mixed well with minerals in the soil.

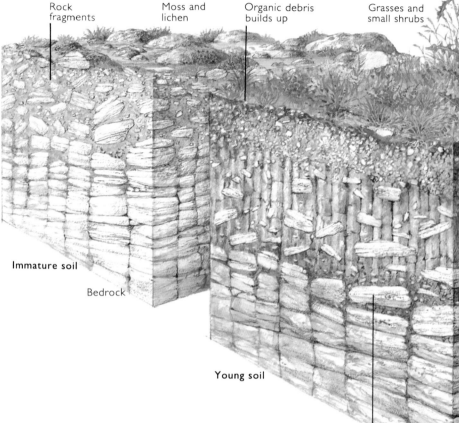

In a mature, well-developed soil, or typical brown earth, the layers are very obvious.
   The thin, dark, topmost layer or O horizon consists of decomposing leaf

Rock fragments

Moss and lichen

Organic debris builds up

Grasses and small shrubs

Immature soil

Bedrock

Young soil

Regolith

In the first stage of soil formation bare rock, either from the bedrock or from a layer of sediment, begins to break up near the surface as a result of both mechanical and chemical weathering.
   Only very tough vegetation, such as lichens and mosses, which can survive on the surface of these rock fragments, is able to take hold. This first organic action further aids the breakdown of the parent material, releasing minerals. Water percolating through the material begins to distribute the minerals, creating an immature soil.

In the next stage the young soil begins to support a greater variety of plants. The lichens and mosses at the surface have provided the organic basis for more advanced plants, such as grasses and small shrubs, to colonize.
   As more plants move in and then die an organic layer develops. Bacteria and microorganisms in the soil act on this organic material, forming humus and releasing nutrients vital to plant growth. Distinct layers, or horizons, begin to be visible in the soil and the break up of the parent material continues.

litter. The richest layer is the A horizon, or topsoil, dominated by the root systems of plants and the organisms of decay. Soil creatures such as millipedes, mites and moles enrich the soil by helping water and air move through it.

The enhanced fertility of the A horizon allows taller shrubs and trees to grow. Humus gives it its dark colour but toward the bottom the soil becomes lighter as water percolating down leaches out soluble substances like organic and mineral salts and metals.

The next layer is the B horizon, or subsoil, where the leached substances are deposited. These give it a distinct colour depending on the mineral content – a brown or reddish hue is due to traces of iron. The particles in this layer are generally coarser than the topsoil. The parent material is found in the lowest level, the C horizon.

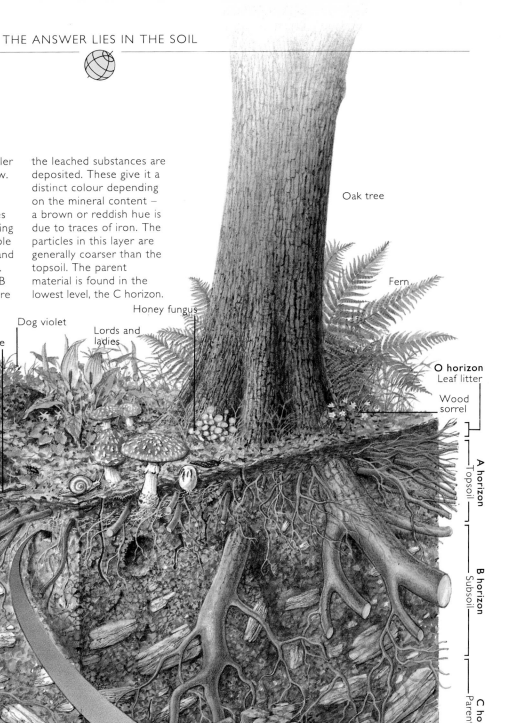

Oak tree

Fern

Honey fungus

Lords and ladies

Dog violet

Millipede

Earthworm

Mole

O horizon
Leaf litter

Wood sorrel

A horizon
Topsoil

B horizon
Subsoil

C horizon
Parent rock

Mite

Nematode

Actinomycetes

Fungus

Bacteria

Mature soil

Root system

Pseudo scorpion

Red earth mite

Springtail

The answer lies in the soil/2

This nutrient-rich layer is often dark and has a fine texture. Next comes the B horizon, or subsoil, where there is much less organic matter and the structure of the soil is coarser. Last comes the C horizon, dominated by regolith which sits on the underlying bedrock. The particles are large and coarse.

The most significant factor in soil development is the climate. Where it is warm and there is a steady, but not excessive, rainfall, soil evolves rapidly. Too little water results in poor humus and too much leads to erosion or waterlogging which starves the soil and plants of oxygen and leaches out nutrients.

In warm areas, chemical reactions in the soil proceed rapidly and the parent material is quickly broken down. Warmth also accelerates the activity of organisms and minerals in the soil. Finally, the temperature controls the rate of evaporation of water by the plants and thus how much water is retained in the soil.

Working inside the soil to improve its structure and fertility are the organisms of the soil. These include everything alive, from tiny bacteria to animals and plants living or growing in the soil. Plants and bacteria play a vital role in organizing the soil, since they release the minerals essential to growth from decaying organic matter. The root systems of plants serve to bind the soil together which helps improve structure and prevents erosion by the elements. Bacteria are particularly active and also assist in the chemical weathering of the regolith.

On a larger scale, animals and insects aid the movement of water and air in the soil as they burrow down, mixing the layers and taking humus down to the lower levels. Earthworms are perhaps the most important of all soil creatures, since they pass both soil and organic matter through their guts, in the process aerating the soil, breaking up leaf litter on the surface and moving the material vertically from the surface to the subsoil. This enhances soil fertility, which in turn provides a rich base for life. So efficient are earthworms that it has been estimated that they turn over the equivalent of all the soil on the planet to a depth of 1 inch (2.5 cm) every 10 years.

A herd of eland
(*Taurotragus oryx*) moves across a landscape dotted with tufty coarse shrubs in the Kalahari Gemsbok National Park in southern Africa. The soils which support the sparse vegetation are a deep terracotta, indicating the presence of iron. Found only in the tropics, this type of soil is termed ferrasol after *ferra,* the Latin for iron.

The region is mainly covered in desert soil which has hardly any humus. This is known as xerosol. Any organic material that may occur, for example from animal droppings or rotting carcasses, is so slight that it merely darkens the soil surface for a short time, but does not allow a horizon to develop.

Flows and slides

Gravity, it is said, is a great leveller. And certainly it is the driving force behind downslope movement of soil, loose stones and rock. But gravity alone does not cause spontaneous falls of material that is in equilibrium. Water is the other crucial factor and it plays a dual role, both reducing the material's cohesive strength and increasing its weight. This explains why slides happen more often in wet weather, and especially in storms. Other factors that can destabilize a slope are vibration – usually from an earthquake – expansion and contraction of the soil, and undercutting by waves or streams.

Material can creep, slide, flow and fall. Creep probably occurs on most hillsides and is a slow, grain by grain movement of material – usually soil – either in the surface layers or deeper. Fluctuations in temperature or moisture cause the expansion or contraction of soil which then creeps downward, virtually imperceptibly or up to $3^1/_2$ inches (9 cm) a year.

Slides travel at between 2 inches (5 cm) per year and a speedy 10 feet (3 m) a second. Most slides are small and shallow at only $6^1/_2$–13 feet (2–4 m) deep and perhaps tens of yards long, but slides in mountainous areas can remove entire hillsides. In translational slides, movement is parallel to the ground, along a straight shear surface. By contrast, in thick homogenous material, the material falls along an arc or semicircular shear plane in a rotational slide.

Flows often form when the material of a slide or fall breaks up and are most common in cohesive materials such as clay, silt and sand. The key feature of a flow is that the material tumbles and rolls turbulently, and does not hold its shape.

Falls are free, or nearly free, movements of rock, or occasionally soil, through the air and are typical on very steep slopes and cliff faces. Most falls start when weathering processes, such as frost action, produce loose rock fragments, which can range from small angular material only inches across to slabs – thick slices of rock several yards across.

Flows and slides induced by humans can be some of the most widespread and devastating slope failures. Farming, deforestation and civil engineering all affect vegetation cover, soil and drainage, sometimes with dire consequences. Deforestation in Madagascar, for example, has resulted in tens of thousands of major failures.

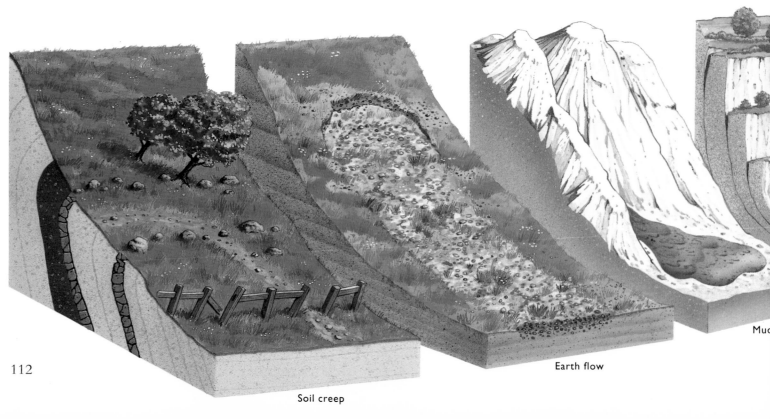

Soil creep

Earth flow

Mud

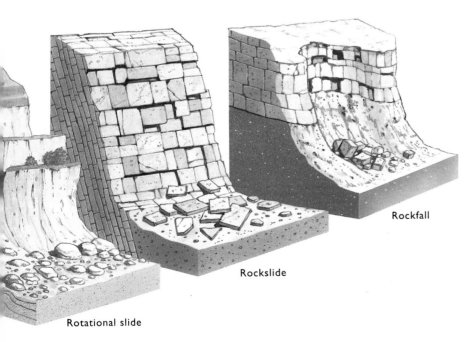

Rotational slide

Rockslide

Rockfall

Soil creep is probably the most common type of downslope movement. Material moves almost imperceptibly, usually in a series of small steps. In a typical earth flow the ground breaks up and tumbles down the slope. It can either be slow, when the moisture content of the soil is low, or fast, when the soil is saturated. In a mudflow, fluid mud, often mixed with rock fragments, moves along a well-defined track and spreads out at the bottom of the slope.

Rotational slides are generally slow but can be catastrophic. A block of earth or rock travels along a curved underlying surface. The block rotates along this curved plane and the toe, or bottom, of the block may disintegrate

to form a debris slide. Rockslides are rapid movements of rock slabs downslope, along bedding planes or joints in the rock. Rock accumulates as an apron of debris at the base of the slope.

Rockfalls are free falls of individual rock fragments down a slope's face. They are very rapid and are common along coastal cliffs. Rock debris builds up a cone at the base.

The wide scree slopes which sweep down below the steep limestone cliffs of Eglwyseg Mountain in the Llangollen Hills in Wales (above) are the result of numerous rockfalls. Weathering of the joints loosens the blocks of limestone which then plummet from the cliff face and crash to the slopes below.

113

Plants and vegetation

Vital both to the planet and to humanity is vegetation, which covers four-fifths of the Earth's land surface. It is easy to assume that the plants that grow in a certain spot have always been there and – if not interfered with by humans – always will be. However, the range of plants in an area is anything but static. New plants are constantly invading and competing with existing plants, and only the fittest survive to grow and reproduce themselves. Changes in an environment tilt the balance in favour of different species.

The change in vegetation over time is succession. Eventually, where conditions in an environment remain more or less stable for a long period, an equilibrium is reached in which both the range and volume of vegetation

If a cultivated field in the mid-latitudes is abandoned to nature the plants that move in to colonize it gradually change in a process called succession (right).

By virtue of its history an agricultural field would generally have good soil cover. As a result colonization starts promptly with the arrival of grasses which become established in the space of a few years.

As these plant pioneers die and decay, organic matter is added to the soil, improving its fertility and allowing dwarf shrubs to colonize the area. These then compete with the grasses for space, light and nutrients. Within a matter of years shrubs have largely taken over, overshadowing the grass. These dwarf shrubs provide the raw materials for taller shrubs and, in turn, young trees which

become established after 25 years or so.

The first trees to grow are pines which are tolerant of the still poor soil. In time, the evergreen pines are replaced by deciduous broadleaf trees such as oak, hickory and maple. These are the ultimate vegetation cover in the mid-latitudes and take over more than 100 years after the field is no longer under cultivation and left to its own devices.

Forest, savanna, desert, grassland and tundra are the world's major biomes – biological communities, each with its own characteristic plant life.

The tropical, or equatorial, zone is dominated by rainforest in areas where the rainfall is high. As it becomes drier, this gives way to scrubland and the widely spaced trees and occasional woodland of savanna. In the driest regions desert prevails. Farther north and south of the equator, in the temperate zone, come the deciduous forests or, in drier areas, chaparral, grassland or prairie.

Nearer the poles, in the subpolar zone, huge tracts of coniferous forest dominate, especially in the north. These finally give way to barren and hostile tundra in the polar zone, where only a few hardy plant species such as lichens, mosses and stunted dwarf birches are able to survive.

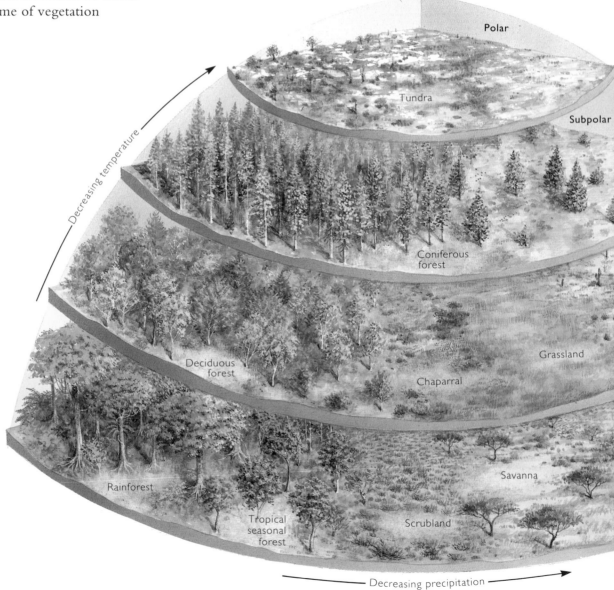

Decreasing temperature

Polar

Subpolar

Tundra

Coniferous forest

Grassland

Deciduous forest

Chaparral

Savanna

Rainforest

Scrubland

Tropical seasonal forest

Decreasing precipitation

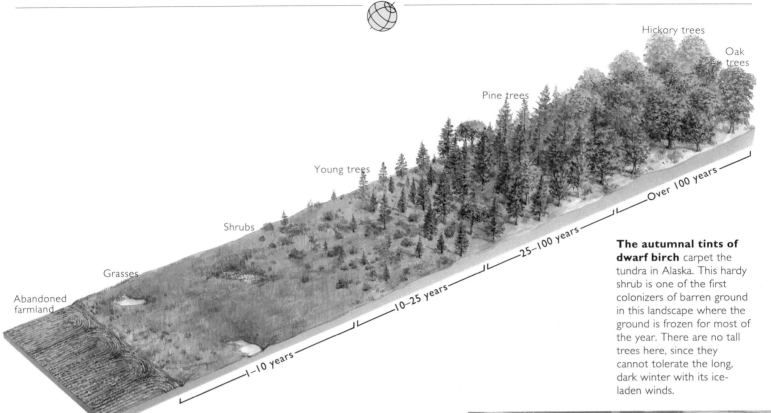

Abandoned farmland

Grasses

Shrubs

Young trees

Pine trees

Hickory trees

Oak trees

1–10 years

10–25 years

25–100 years

Over 100 years

Temperate

Desert

Tropical

Desert

Equator

The autumnal tints of dwarf birch carpet the tundra in Alaska. This hardy shrub is one of the first colonizers of barren ground in this landscape where the ground is frozen for most of the year. There are no tall trees here, since they cannot tolerate the long, dark winter with its ice-laden winds.

increase to the maximum that the environment can support, whether it be sparse, dry desert or rich, wet rainforest. This steady state is known as climax vegetation as it is the end point of the succession of the plants in the area.

Plant colonization of a new, bare surface is primary succession. In the early stages, tough, adaptable plants such as lichens and mosses predominate. They encourage soil development and release nutrients which enable other plants to move in. Next come grasses and herbs, followed by shrubs and, finally, trees.

The main form of succession is secondary succession, which happens when fire, drought or farming destroy or remove vegetation. At each stage a new set of conditions is established, and as a result the previous vegetation is eliminated. As the sequence progresses, the soil develops in parallel with the vegetation: more organic matter accumulates and the soil can hold more water.

A microclimate is created where the vegetation has a softening effect on the local climate and environment. As vegetation develops its root systems stabilize the soil and reduce erosion. Conversely, where well-established vegetation is cleared away, rainfall can remove the soil and in due course the landscape will change. This is happening to devastating effect in tropical rainforest areas where logging is rife.

BEYOND
THE LAND

Beyond the land

Dynamic planet Earth is vibrant and alive with billowing clouds, rushing winds, spiralling storms, rippling waves and swirling ocean currents. The oceans and atmosphere, together with the continents, constitute the life-sustaining layer of our planet – a vital layer but one that is surprisingly thin in relation to the function it fulfills. Shrink planet Earth, with a diameter of nearly 8,000 miles (12,800 km), to the size of a large apple and the oceans and weather atmosphere would be thinner than the apple skin.

The oceans and atmosphere are intimately coupled by energy and water cycles that operate on a global scale. About half the incoming sunlight is absorbed by the planet's surface, especially the tropical oceans. This warms the sea and generates evaporation and rising air currents which transfer heat and water vapour from the oceans to the atmosphere. When the water vapour condenses into clouds the process releases energy which causes the rising air currents to surge to the top of the weather atmosphere and then begin to sweep poleward.

The transfer of energy from the tropical oceans to the atmosphere signals the start of a circulation pattern of winds that encompasses the globe. This is responsible for the different types of regional weather and climate worldwide and also determines the pattern of surface ocean currents.

The oceans and the atmosphere work in tandem to redistribute heat between the intensely hot equatorial regions and the frigid poles, thus ensuring that low latitudes are not overheated nor the high latitudes supercooled. At sea, surface ocean currents driven by the winds carry warm water to the poles and cold water to the equator. In the atmosphere warm and cold winds interact to exchange heat, and moist winds release heat when the water vapour forms clouds.

Storms play a not insignificant role in energy transfer systems, especially the cyclones of the tropics and mid-latitudes. Tropical cyclones, or hurricanes, are like the safety valves of an overheating boiler, brought into operation when and where the tropical oceans become explosively hot. The phenomenal transfer of heat and moisture from the ocean to a hurricane is reflected in the incredible wind speeds around the eye of the hurricane, as well as the towering clouds and torrential rain generated by the storm.

About 100 of these large circular storms develop each year, especially in the Pacific Ocean, where they are called typhoons. They move west with the trade winds, wreaking untold devastation on coastal communities with powerful storm surges, deluges of rain, and winds gusting up to 250 mph (400 km/h) and spawning tornadoes. Leaving chaos in their wake they finally turn into the mid-latitudes and decay.

Through mid-latitude cyclones, also known as frontal depressions, energy is exchanged across the air temperature boundary between middle and polar latitudes. This fluctuating boundary, called the polar front, marks the combat zone between warm moist

Like a floating blanket tinged red by the setting Sun, altocumulus clouds stretch toward the horizon (previous page). The appearance of these clouds in the sky often indicates the impending approach of rain.

winds, or air masses, of tropical origin and cold dry winds flowing out from the poles. Without frontal depressions these opposing sets of winds would clash – but would not mix – and be sent aloft to go back to where they started from without having achieved the exchange of energy.

A number of frontal depressions assemble along the front, and the air masses within each storm are swirled around each other. Over several days warm air is whirled to the pole side of the storm and cold air to the equator side to be absorbed into the surroundings, achieving the vital energy transfer. The birth, growth and death of a procession of these storms is orchestrated from above by a narrow band of high-altitude powerful winds called a jet stream.

J ust as the atmosphere fashions a global circulation pattern so the surface currents and deep currents of each ocean link up in a worldwide circuit. This conveyor belt is driven by the unequal sizes of the oceans since the Pacific Ocean gains more heat from the Sun than the Atlantic Ocean.

Cold salty waters around Iceland in the North Atlantic sink and flow south as a deep current, merging with the deep cold waters around Antarctica before flowing into the North Pacific. There they release their salty load, become less dense and rise to the surface, returning to the North Atlantic with considerable warmth from the Pacific.

This circuit keeps the North Atlantic much warmer than it would otherwise be and maintains a heat balance between the world's oceans. This warmth is transferred to the westerly winds and Europe enjoys the benefits by experiencing a mild climate.

The oceans and atmosphere are linked in many other ways. Every few years in the Pacific Ocean, a sudden surge of warm surface waters from the central Pacific toward the eastern Pacific leads to a complete reversal of the trade winds. Whereas the trades normally blow westward, creating clouds and heavy rainfall over Indonesia and northern Australia, their sudden reversal during these El Niño periods induces dry conditions, even severe drought, in these regions. The consequences for the eastern Pacific are equally dramatic as the eastward blowing moist trade winds cause damaging floods and mudslides on the

normally dry west coast of South America.

Winds, storms and ocean currents interact with one another to create distinctive climate zones, with places in each zone experiencing a similar seasonal cycle of weather. These zones have a fundamental influence on the agriculture, ecosystems, water supplies, human comfort and weathering processes that shape the landscape.

However, within each region there are considerable local variations in weather and climate because of the influence of individual features such as hills, valleys, lakes and seas. It is this infinite variety of weather, as well as its changeability and, all too often, its unpredictability, which challenges us to continue to explore the nature and workings of the atmosphere and its links with the oceans.

Frozen seas

Never-ending landscapes of snow, ice and water characterize the polar latitudes. Only slowly do these frozen wilderness scenes change as the thin broken sea ice expands and contracts around the poles, and icebergs break off, or calve, into the sea from coastal glaciers and drift with the ocean currents.

The North Pole is centred on the Arctic Ocean with its thin covering of permanent sea ice, whereas the South Pole lies amid ice-covered Antarctica. This continent, twice the size of Australia, is almost entirely blanketed by ice sheets with an average thickness of 6,500 feet (2,000 m). If all this ice melted it would raise world sea level by 180 feet (55 m).

Sea ice forms during the long, dark and bitterly cold polar winters. The water surface is chilled and freezes into ice which floats because it forms a larger volume than a similar mass of water and so is less dense. Most sea ice is less than 6½ feet (2 m) thick but areas of permanent Arctic sea ice can build up to 10-16 feet (3-5 m).

Icebergs form when blocks of freshwater ice break from the snouts of glaciers reaching the coast. Glaciers along the coasts of Greenland, Labrador and Antarctica produce tens of thousands of icebergs annually. Icebergs from Greenland and Labrador travel southward in the Labrador Current and gather off the Newfoundland Grand Banks. There they meet the warm Gulf Stream, break up and melt. Since only one-eighth of an iceberg is visible above the sea surface, this poses a danger to shipping. The *Titanic* discovered this on her maiden voyage in 1912, when she struck an iceberg and sank off the coast of Newfoundland with the loss of 1,517 lives.

Extensive lobes of ice sometimes spill out over the sea creating a floating ice shelf. The Ross Ice Shelf of Antarctica, the size of France, produces huge flat-topped icebergs in contrast to the smaller, irregular shapes generated by glaciers. In 1991, a massive iceberg 75 miles (120 km) long detached itself from the Weddell Sea Ice Shelf and drifted northeast at a rate of 9 miles (14 km) a day, crossing the shipping lanes between Antarctica and Argentina.

Maximum extent of winter snow and ice in the polar latitudes (below). Sea ice trapped in the heart of the Arctic Ocean drifts slowly clockwise taking about a decade to complete a circuit. Sea ice now covers 10 percent of the world oceans but during the Little Ice Age (1450–1850) sea ice enveloped Iceland and expanded around Greenland, preventing Norse voyages to North America and isolating their Greenland colonies.

Glaciers and ice sheets cover 10 percent of the world's land area today, but at the height of the last glaciation, 18,000 years ago, covered 30 percent, with thick ice sheets over the northern parts of North America, Europe and Asia. By contrast, four million years ago even Greenland was ice free.

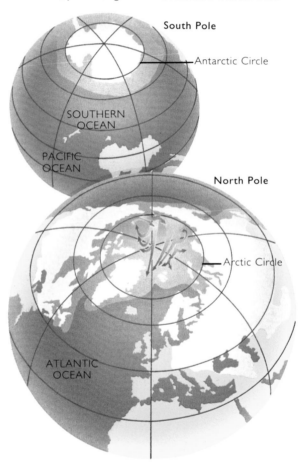

South Pole

Antarctic Circle

SOUTHERN OCEAN

PACIFIC OCEAN

North Pole

Arctic Circle

ATLANTIC OCEAN

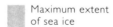

Maximum extent of sea ice

Maximum extent of snow and ice on land

Pattern of ice drift

Sea ice in Lancaster Sound in the Canadian Arctic. This channel, ice free for a few months in summer, is one of several forming the North West Passage from the Atlantic to the Pacific. For centuries, this route was sought by explorers and many lost their lives in the search.

Dynamics of the seas

The swirling eddies of surface currents that stir every ocean on Earth fit together to create an intricate global network of ocean currents. Drop a message in a bottle off Australia and, several years later, it may be washed up on a beach in Florida or Scotland.

Like the global winds that generate and sustain them, the planetary role of currents is to ensure the tropics do not overheat. Consequently, surface waters warmed in the tropics swirl poleward, while waters cooled at high latitudes flow toward the equator.

Ocean currents exert strong influences on the climate of neighbouring continents. The warm Gulf Stream ensures northwest Europe enjoys a mild climate for its latitude, keeping Norwegian and Icelandic harbours ice free. In contrast, the cold Peru, or Humboldt, Current causes onshore winds in Chile to be cold and dry. This maintains the aridity of the Atacama Desert.

Each major ocean basin displays an immense oval-shaped flow, or gyre, of ocean currents. The surface currents of each oceanic gyre are driven by the overlying trade winds and westerlies, and the currents on the eastern and western sides of the gyres are strongly influenced by the shapes of the continents and the direction of rotation of the planet.

The currents flowing along the western side of northern hemisphere gyres (western boundary currents) are fast, deep and narrow.

Those on the eastern side (eastern boundary currents) are slower, wider and shallower.

In the North Atlantic the western boundary current is the Gulf Stream and its counterpart in the North Pacific is the Kuroshio. Both are only about 60 miles (100 km) across and flow at speeds greater than 6 feet (1.8 m) per second. The Canaries Current and the California Current, however, are more than 600 miles (1,000 km) wide and move at less than 1 foot (30 cm) per second.

In the southern hemisphere, the differences between boundary currents are less marked because the eastern boundary currents of the South Atlantic and Pacific oceans are strengthened by the strong Antarctic Circumpolar Current (West Wind Drift). This powerful current, the only one that flows all the way around the world, is driven around Antarctica by the westerly winds of the Roaring Forties and Fifties.

Surface currents undergo slight seasonal changes in both strength and location in response to changes in wind and climate patterns. The current just off the coast of India even undergoes changes in direction because the overlying winds reverse direction in accordance with the Southeast Asian monsoon. In the summer the winds drive the current eastward, and in the winter they blow it in the opposite direction westward.

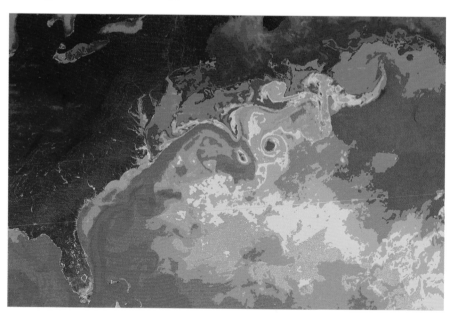

The Gulf Stream, the western boundary current of the North Atlantic, is seen here in a satellite photograph colour-coded by computer to show warm water in red and cooler water in yellow. Drifting alongside the Gulf Stream are eddies of warm water to the north and cold water to the south, each 60–190 miles (100–300 km) in diameter. At any one time as much as 40 percent of the continental shelf of the North American east coast may be covered by warm eddies and 15 percent of the Sargasso Sea covered by cold eddies.

Indian Counter Cu

North Equatorial Cur

South Equatorial Cu

Agulhas Current

IND

OC

West Austral

➤ Warm currents

➤ Cold currents

The surface currents of the oceans, extending to depths of several hundred feet, are controlled by the global wind pattern. Currents around the ocean basin (gyres) flow clockwise in the northern hemisphere, counterclockwise in the southern hemisphere. The water at the centre of a gyre is relatively slow-moving because there are no strong currents there. This makes them ideal breeding grounds for marine species, notably the eels of the Sargasso Sea of the North Atlantic.

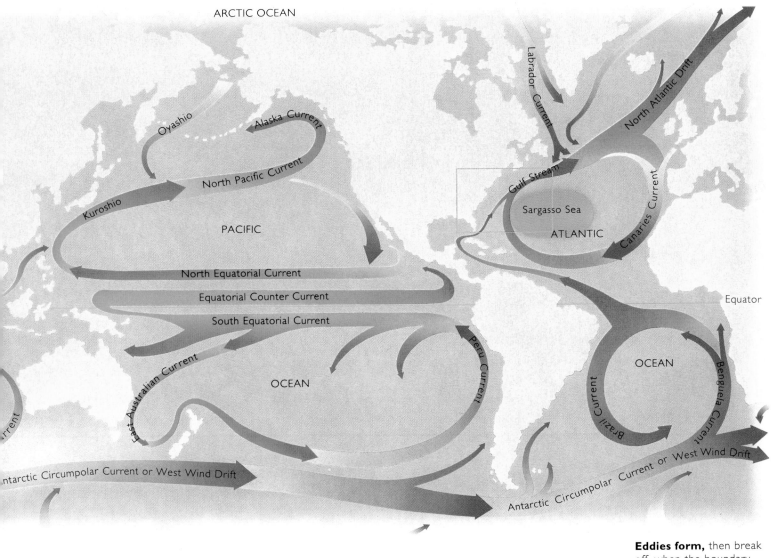

ARCTIC OCEAN

Oyashio

Alaska Current

North Pacific Current

Kuroshio

PACIFIC

North Equatorial Current

Equatorial Counter Current

South Equatorial Current

East Australian Current

OCEAN

Labrador Current

North Atlantic Drift

Gulf Stream

Sargasso Sea

ATLANTIC

Canaries Current

Equator

Peru Current

Brazil Current

OCEAN

Benguela Current

Antarctic Circumpolar Current or West Wind Drift

Antarctic Circumpolar Current or West Wind Drift

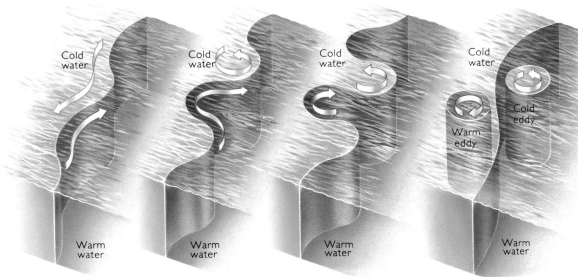

Cold water

Cold water

Cold water

Cold water

Cold eddy

Warm eddy

Warm water

Warm water

Warm water

Warm water

Eddies form, then break off, when the boundary between opposing flows of warm and cold water wavers. This often happens shortly after severe weather storms have passed over the boundary. Eddies retain their physical identity for months or even years and fishing fleets track the deep, productive, nutrient-rich cold ones using satellite photography.

Occasionally, eddies, nicknamed "Meddies", form from the salty water spilling out of the Mediterranean Sea through the Strait of Gibraltar.

123

Dynamics of the seas/2

Surface and deep-sea ocean currents join together to form a global conveyor belt. Surface waters of the North Atlantic provide the starting point by cooling, sinking, and sending a cold dense current southward beneath the Gulf Stream eventually to join the cold dense waters formed near Antarctica.

This combined deep-sea current then branches into the Indian and Pacific oceans before rising to the surface and eventually returning some of the warmth from those oceans to the North Atlantic.

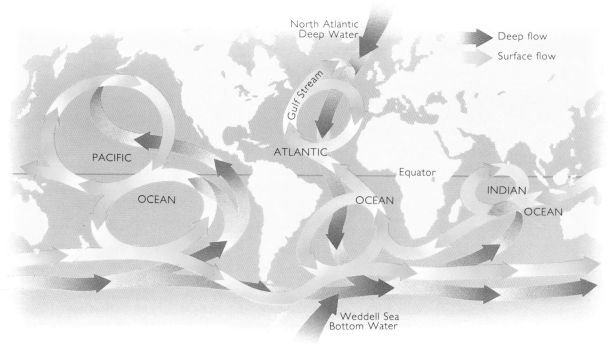

Deep beneath the sea lie powerful but very slow-moving currents which join with surface currents to form a vast oceanic network of circulating water, transferring energy, nutrients and sediments around the world. Deep-sea rivers flow over spectacular sea-floor landscapes of towering mountains, seemingly bottomless trenches, and extensive plains.

Deep-sea currents, each carrying more than 20 times the amount of water transported by all the world's rivers combined, are generated when surface waters increase in density, which makes them sink and flow along the ocean floor. During the long polar winters, waters in these regions cool rapidly. As surface water temperature falls, its density increases and it sinks, creating the deep-sea currents. Sea ice adds to the density of the water beneath by ridding itself of salts during its formation, thus leaving the underlying water more saline.

Surrounding the continents of the world, beneath the offshore waves, lies a gently sloping continental shelf. This coastal apron, with its seaward edge at an average depth of 425 feet (130 m), once formed part of the continental landmasses. It covers a substantial area but varies in width, being barely evident along the west coast of South America but extending over 1,000 miles (1,600 km) into

the Arctic Ocean from the Siberian coast.

Much of the shelf was formed by wave action when world sea level was 300–400 feet (90–120 m) lower than it is today. Sea level has fluctuated 10–15 times in the past two million years. At times during this period thick ice sheets covered much of middle and high latitudes of the northern hemisphere, trapping water evaporated from the oceans in the ice. As a result, many areas of the continental shelf reveal mounds and ridges of glacial debris, deeply gouged U-shaped valleys, lofty volcanoes, enormous deltas, and tall coral reefs.

Deep, steep-sided canyons, similar in size to the Grand Canyon of the United States, cut through the continental shelf. The Congo Canyon, off the west coast of Africa, is 6 miles (10 km) wide at its top and 2 miles (3 km) deep. Of similar depth is the Hudson River Canyon which extends 80 miles (128 km) seaward from the mouth of the Hudson River.

Although some canyons are simply seaward extensions of river valleys, others are not. The first clue to their origin came in 1929 when an earthquake shook the Newfoundland coast and 13 transatlantic cables were severed. Initially it was thought the tremor had caused the breaks but the timing and distance between each break suggested otherwise.

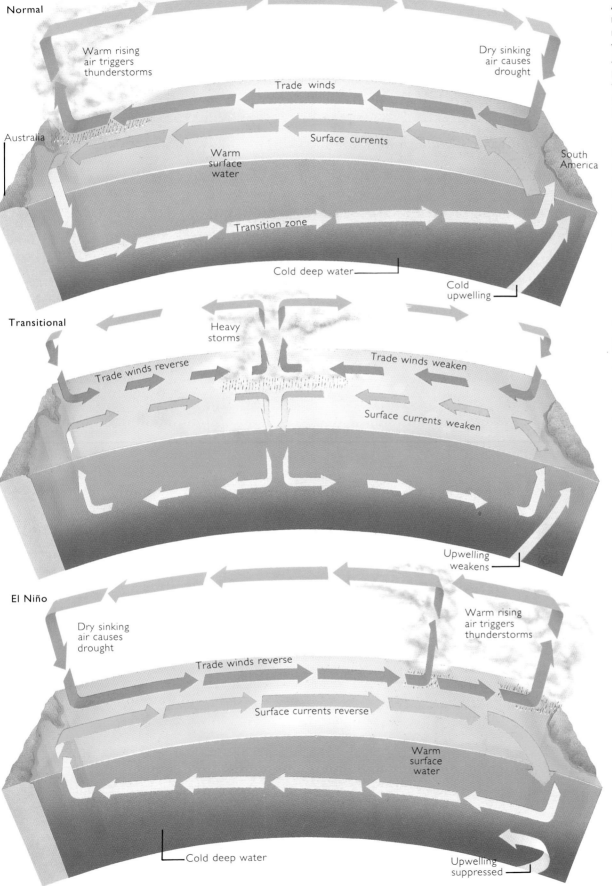

Normal

Warm rising air triggers thunderstorms

Dry sinking air causes drought

Trade winds

Australia

Surface currents

Warm surface water

South America

Transition zone

Cold deep water

Cold upwelling

Transitional

Heavy storms

Trade winds reverse

Trade winds weaken

Surface currents weaken

Upwelling weakens

El Niño

Dry sinking air causes drought

Warm rising air triggers thunderstorms

Trade winds reverse

Surface currents reverse

Warm surface water

Cold deep water

Upwelling suppressed

An upwelling of cold nutrient-rich water normally occurs off the west coast of South America. Trade winds push warm surface water westward where it piles up, raising the sea level by 16 inches (40 cm).

Air above the cold upwelling sinks, producing dry weather conditions. Air above the pool of warm waters forms rainfall over Indonesia and Australia. If the trades are strong, very heavy rainfall occurs while drought affects South America.

The trade winds change direction every few years. The warm surface waters accumulated in the west move eastward, pooling in the central Pacific and beginning El Niño (Spanish for boy). Heavy rains and storms affect the central Pacific and droughts begin in Australia and Indonesia.

El Niño plays havoc with the world's weather, worsening drought in the Sahel, weakening the Indian monsoon, pushing mid-latitude storms poleward and causing drought in the central U.S.

The warm waters reach the Pacific coast and suppress the upwellings of cold water. Fish die without this nutrient-rich water and the fishing industry collapses. The marine food chain is disrupted and marine mammals and seabirds starve.

The warm surface waters off South America trigger rainstorms, floods and mudslides along the normally dry Pacific coast, while severe drought, heatwaves and bushfires occur in Australia.

Dynamics of the seas/3

In fact, what had caused the cables to break was an avalanche of sediment-choked water called a turbidity current. It was triggered by the earthquake and raced down the continental slope, snapping the cables.

The sunlit waters above the shallow continental shelves support a rich variety of marine species. In the deeper ocean beyond the shelf, the absence of sunlight means plants cannot survive and creatures need special adaptations to thrive in the dim light – or even complete darkness – and to withstand the intense water pressure and numbing cold.

Our knowledge of life in the deep oceans is incomplete since this remains one of the most difficult parts of the world to explore. One recent discovery, off the coast of Madagascar, was living specimens of coelacanth, a species related to freshwater lungfish. This "living fossil" fish, an ancestor of the first land vertebrates, was previously assumed to have been extinct for 60 million years.

Most of the world's ocean floors are carpeted by a deep layer of loose sediments, reaching a thickness of 1,640–3,300 feet (500–1,000 m) in the Atlantic and Pacific oceans. In deep ocean trenches, which readily trap the sediments, accumulations exceed 5 miles (8 km). Some sediments are formed by the tiny rock fragments or mineral grains weathered from the continents and transported to the oceans by rivers, winds and ice. These tend to be red or brown in colour because iron on the particles reacts with dissolved oxygen in the water to produce a coating of rust.

Other sediments are formed from the billions of minute shells and skeletons of marine animals and plants that are found in surface waters and which dissolve as they sink into the deeper cold waters. These sediments cover vast areas of the ocean floor with a thick, soft calcareous ooze.

Some ocean sediments consist of minerals that crystallize from sea water through chemical reactions. They include limestone-type deposits and manganese nodules. The latter are black, layered lumps, formed in a slow chemical reaction, taking 1,000 years to produce a rounded or flattened specimen only a few inches across. They can contain 10–30 percent manganese, as well as significant quantities of cobalt, copper, iron and nickel, and will prove a valuable resource when the technical problems of removing large quantities from the deep oceans have been overcome.

Soft corals and sponges adorn the shallow waters of the Caribbean Sea. These creatures thrive on the calcareous reefs built by stony corals and coralline algae.

The delicate, finely branched corals and tubular sponges inhabit the more protected areas of the reef. They can only survive in warm clean seas and live close to the surface.

Reefs are the largest biological constructions on Earth and are second only to tropical rainforests in their diversity of species.

A gently sloping continental shelf gives way abruptly to a steeply shelving continental slope and less steep continental rise before reaching an extensive abyssal plain at a depth of 2 1/2 miles (4 km).

Deep submarine canyons are cut into the continental slope as a result of a sudden cascade of dense, mud-choked water (turbidity current) rushing down its slope at peak speeds of 50 mph (80 km/h). These muddy torrents can reach a vast thickness as loose sediment on the shelf and slope is dislodged and thrown into suspension by earth tremors – similar to the effect of throwing a rock into a lake with a muddy bed.

Some turbidity currents have been followed from the North American continental shelf to the Mid-Atlantic Ridge, but most travel less than 60 miles (100 km) before stopping.

Turbidity currents transport fragments of plants and animals that live only in the shallow continental shelves and this explains why such remains are sometimes found in sediment cores taken from the deep oceans.

Where land and sea meet

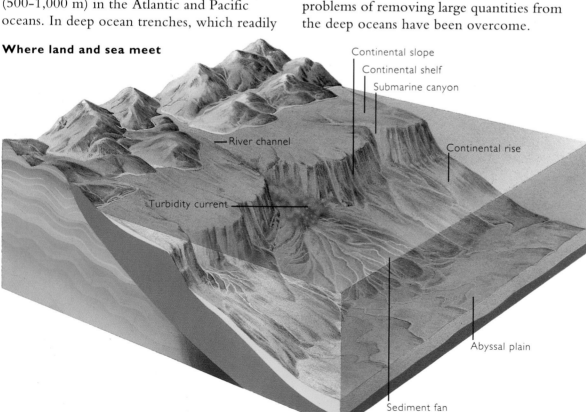

Continental slope
Continental shelf
Submarine canyon
River channel
Continental rise
Turbidity current
Abyssal plain
Sediment fan

Rhythms of the tides

The world's oceans are in constant motion being stirred by powerful currents, waves and tides. Unlike ocean currents and waves, which are driven by the winds and derive their energy ultimately from the Sun, tides are created by the gravitational attraction forces of the Moon and, to a lesser extent, the Sun.

The Earth and Moon rotate once every 28-day lunar month around their common centre of mass which lies close to the centre of the Earth. This eccentric rotation means that the Moon's gravitational force raises an oceanic bulge on the side of the Earth facing the Moon, and an opposing centrifugal force raises a similar oceanic bulge on the other side. The Earth rotates, sweeping beneath these two bulges and their intervening troughs, creating the daily rise and fall of the oceanic tides.

In the open sea the tidal range may be only 3 feet (l m) but as a tidal bulge moves over the shallow continental shelves, or through narrow straits, or into estuaries the tidal range is amplified. By contrast, enclosed seas such as the Baltic and Mediterranean, and even the Caribbean to some extent, are almost tideless.

In narrow estuaries with large tidal ranges, the body of water rushing in with the tide is concentrated so much that it builds into a rolling wall of water, or tidal bore. Tidal bores can reach a height of 5 feet (1.5 m) in the Severn Estuary in England, and 26 feet (8 m) on the spring tides of the Tsientang River of northern China.

Tides have a strong influence on the vertical distribution of plants and animals along the coast. Along rocky shorelines, the common periwinkle snail (Littorina) occupies the highest rocks that are covered only at high tide. Below lies the rock barnacle (Balanus) and beneath this zone is found the mussel (Mytilus), which can survive exposure to the air for only relatively short times.

Waves are created when wind blows over a water surface, piling it up into ridges which then move in the direction of the wind. Waves deceive the eye, for it is energy that is being moved, not water. The greater the distance over which the wind blows, as well as the stronger the wind, and the longer the wind has been blowing, the larger the waves. Once formed, waves can travel far beyond the windy area in which they were generated.

When a wave arrives in shallow water, the rising sea bottom causes the base, but not the crest, to slow down. The crest rolls forward, overshoots, and causes the wave to break.

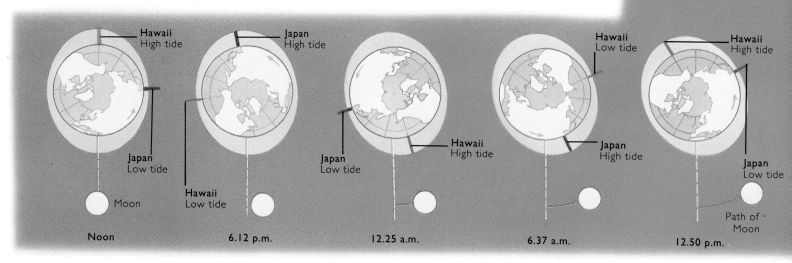

The gravitational pull of the Moon on the Earth results in two bulges in the Earth's oceans, one on the moonside and one on the opposite side of the Earth. As the Earth rotates through its 24-hour daily cycle, a coastal location progresses through the two tidal bulges and two troughs, so experiencing two high tides and two low tides each day.

The complete tidal cycle of two high tides at a location lasts slightly longer than a day – 24 hours 50 minutes. This is because the Earth rotates and the Moon orbits in the same direction and the Earth must rotate another 12 degrees to bring it back in line with the Moon.

One complication in predicting the precise times of high and low tides for a location arises because it takes time for the gravitational effect to move the water. This results in a distinct time lag between the lunar alignment and high tide. The lag is also affected by the shape of the coastline and water depth.

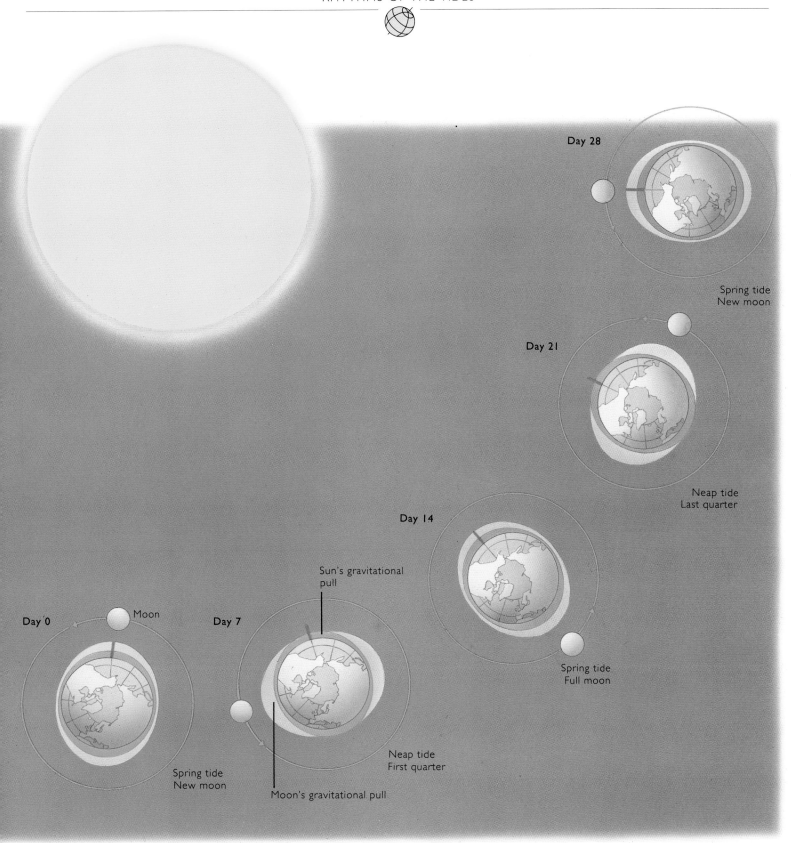

Day 28

Spring tide
New moon

Day 21

Neap tide
Last quarter

Day 14

Sun's gravitational
pull

Spring tide
Full moon

Day 0

Moon

Day 7

Spring tide
New moon

Neap tide
First quarter

Moon's gravitational pull

The tides follow a 28-day cycle reflecting changes in the alignments of the Earth, Moon and Sun and the relative strengths of the gravitational attractions between these bodies. The largest effect occurs when the Sun and Moon are in line, producing spring tides.

Spring tides have the largest tidal range — the highest high tides and the lowest low tides. When the Sun and Moon are at right angles the weaker neap tides are produced, with their small tidal range — the lowest high tides and the highest low tides. Spring tides are about one-third bigger than neap tides.

Additional variances occur in tidal cycles because of the elliptical orbits of the Moon and Earth. When the Moon is closest to the Earth, tides are 20 percent higher. This occurs twice a month. The Earth is closer to the Sun in its elliptical orbit during the northern hemisphere winter. This means winter tidal ranges are greater in the northern hemisphere during winter than they are in summer.

Rhythms of the tides/2

Waves may break several times before washing up the beach as surf.

A storm over the deep ocean creates a spectrum of different waves which travel outward from the storm centre as a group. By the time they have travelled 1,000 miles (1,600 km) or more they have become well sorted. The faster waves leading the group are those with long parallel crests and uniform wavelengths, called swell. Thus when swell reaches the coast it usually foretells an approaching storm.

Tides and waves can be harnessed to produce electricity. If a line of reversible turbines is built across an estuary, the inflowing and outflowing tides can be used to drive the turbines. The first tidal power station was built in 1966 in France to utilize a tidal range of 44 feet (13.5 m).

Harnessing the vertical motion of waves to drive an electricity generator has been pursued for many years with limited commercial success. Apart from the technical difficulties, progress has been hampered by uncertainties over the effects of large-scale interference with waves on the ecology of nearby coastal environments.

A Hawaiian plunging breaker (right). As an incoming wave encounters the rising slope of the shore, it curves over, trapping the air below it as it plunges toward its base. The plunging crest moves along the length of the wave as it approaches the beach creating a tube.

Waves often approach the coast at an angle and then change direction on entering shallow water, reaching the beach parallel to the coastline.

The part of the wave still in deep water travels faster than the part that has reached shallower water. This causes the crest to bend or curve around in a position parallel to the contours of the shore. Thus waves tend to converge on headlands, accelerating erosion there, but spread out in bays, reducing their potential for destruction.

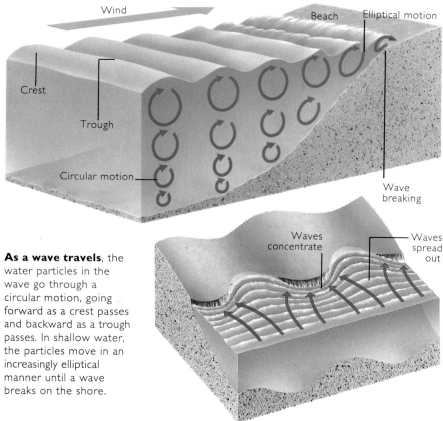

As a wave travels, the water particles in the wave go through a circular motion, going forward as a crest passes and backward as a trough passes. In shallow water, the particles move in an increasingly elliptical manner until a wave breaks on the shore.

Changing coastlines

At the height of the last glaciation around 18,000 years ago, so much water was locked up in massive ice sheets that world sea level was 390 feet (120 m) lower than it is now. As global climate warmed, the ice sheets melted and sea level rose. Continental shelves were inundated, new islands and inlets were created including the British Isles and the Baltic Sea, and new coastlines were attacked everywhere by the forces of the rising sea.

Sea level reached its present level about 5,000 years ago. Since then coasts have been exposed to the relentless pounding of waves, eroding some coasts and causing them to retreat. Other coasts have benefited from the spoils of battle, extending seaward as sediments were deposited to form beaches, spits and bars.

The changing shape and form of coasts reflect the interplay between marine processes and many factors. Whether the rock is hard or soft, the direction in which rock strata dip, and whether the rock contains weaknesses such as joints and faults are important.

The topography of the land affects the shape of the coastline. If rivers and glaciers have carved direct routes through to the sea the coastline is one of long peninsulas and deep inlets aligned at right angles to the coast – the Atlantic or ria type – such as in southwest Ireland. If valleys and ridges lie parallel to the coast, elongated islands and inlets predominate, the Dalmatian or Pacific type of coast.

Given sufficient time, marine processes will create smooth coastlines as headlands are eroded and bays infilled by sediments. Many coasts have not achieved this state of equilibrium because sea level rose to its present height only about 5,000 years ago.

Where coastal rocks are susceptible to the battering of waves, strong erosion will carve tall cliffs and wide shore platforms resulting in rapid coastal retreat. Waves exploit weaknesses within the cliff, resulting in the formation of caves, arches and stacks.

The material eroded from the cliffs, together with sediments swept onshore from the seabed and dumped by rivers, moves along the shore to be deposited as beaches, spits, bars and tombolos, often with sand dunes and salt marshes behind.

Where wave attack is modest and rocks are resistant, erosion is weak and the headlands are only slowly truncated. Spits and bars can develop across the estuaries creating lagoons which infill and help to straighten the coastline.

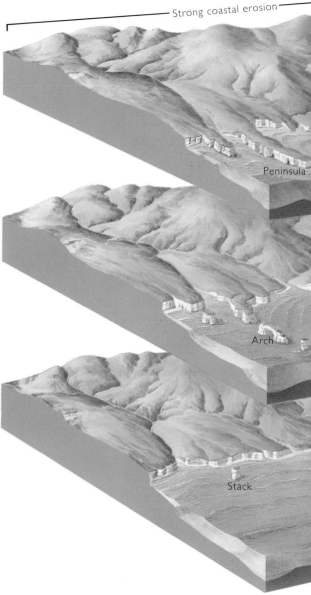

Strong coastal erosion

Peninsula

Arch

Stack

Some fissures in a cliff face can be widened and deepened by wave attack to form caves. Caves then enlarge through roof collapse, triggered by the explosive effects of air being trapped inside them and compressed by the tremendous force of the waves.

When caves open up either side of a headland, a sea arch is created. When the arch lintel collapses a sea stack is isolated from the headland, as here at Cabo San Lucas on the Baja California peninsula.

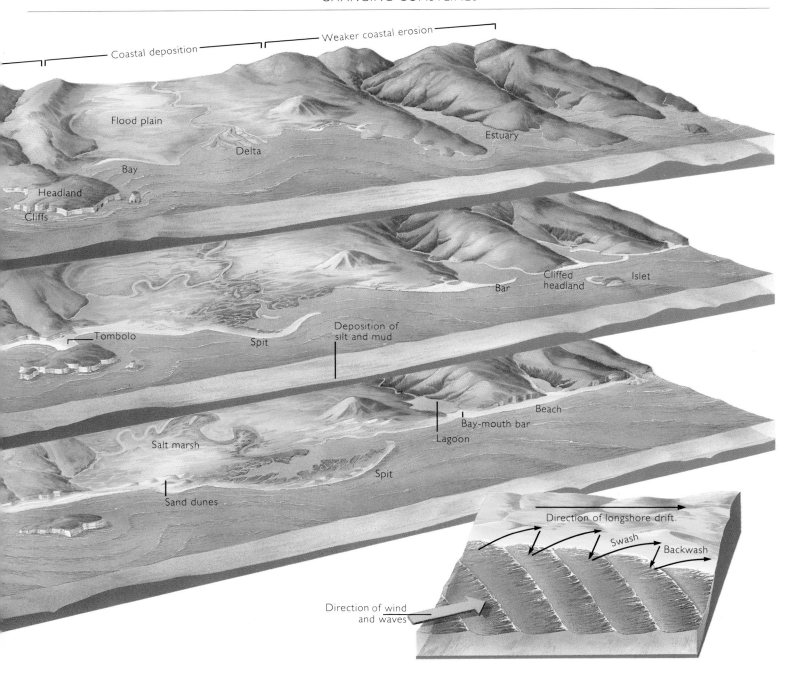

Coastal deposition

Weaker coastal erosion

Flood plain

Delta

Bay

Headland

Cliffs

Estuary

Tombolo

Spit

Deposition of
silt and mud

Bar

Cliffed
headland

Islet

Beach

Salt marsh

Bay-mouth bar

Lagoon

Spit

Sand dunes

Direction of longshore drift

Swash

Backwash

Direction of wind
and waves

Waves assault coasts in many ways. When the mass of water from a powerful wave strikes a cliff its face is subjected to a huge force. Shock waves open up an extensive network of fine cracks. Weaknesses such as joints and faults are exploited as waves compress air in fissures, resulting in explosive blasts which break tiny fragments from the rock.

The fissures deepen and this can dislodge blocks, sending them crashing to the base of the cliff. A narrow wave-cut notch forms at the cliff base which later enlarges into a horizontal shore platform. Undercutting of the cliff causes rock falls and the shore platform is covered by shattered and dislodged material. This is broken down into pebbles forming

missiles for storm waves to hurl at the cliff face. Pebbles and sand are scraped along the cliff base and used to scour the shore platform.

Not all coasts suffered submergence as sea level rose at the end of the last ice age. Some coasts are emerging from the sea as a result of uplift due to volcanic and earthquake activity. Many northern lands experienced rebound after being freed from the colossal weight of ice sheets. Northern Scandinavia, once covered by an ice sheet up to 8,200 feet (2,500 m) thick, is emerging from the sea by as much as 1/3 inch (1 cm) per year.

Where the land rebounded faster than sea level rose, a sequence of raised shore platforms arises, which forms broad steps around the coast.

Where a wave breaks obliquely against a steep beach, its uprush, or swash, drives sand and pebbles diagonally up the beach. The return flow, or backwash, affected by gravity, drags the sediment down the beach at right angles to the shoreline.

Successive waves move the sediment along in a zigzag pattern. Along shores with an irregular seabed, height differences in offshore waves cause water to move from high to low areas creating a longshore current, dragging sediment along as longshore drift.

Changing coastlines/2

These are convenient flat areas for roads and houses, but look strange with their inland cliffs, caves and sea stacks. Such features are evident in parts of west Scotland, the Baltic coast and the Arctic coasts of North America.

Coasts that are advancing seaward are often gently sloping, since large waves break well offshore where they pile up sand to form an offshore bar. These offshore bars, parallel to the coastline, can develop into barrier islands, often covered by sand dunes. The shallow water between the bar and the coast is cut off from the open sea and forms a lagoon.

Over time the lagoon fills with mud and silt from inflowing streams, creating salt marshes which can be drained for use as farmland. Barrier islands may lie many miles from the coast such as those off the northern Netherlands coast. Barrier coasts comprise one-eighth of the world coastline, including eastern United States and the Gulf of Mexico.

To protect coastlands threatened by erosion, massive and expensive vertical and curved-lip sea walls are built. They often crack and fail during severe storms because the sea wall concentrates wave energy against its face rather than dissipating it. The waves reflected seaward by the wall can cause excessive scouring of the beach, exposing the wall foundations. The increased water depth enables even larger waves to reach the wall. Dumping large angular rocks called rip-rap in front of a sea wall offers extra protection as this absorbs the wave energy.

Thoughtless removal of beach material or offshore sediments for use in construction can accelerate coastal erosion. These sands and pebbles were largely produced by river and glacier erosion during the last ice age and deposited on the land that is now covered by the sea. The rising sea scoured these loose sediments, pushing some ashore to form beaches and moulding the rest into offshore bars. Any loss of beach material due to drifting along the coast is replenished by storms transferring sediments from the offshore bars. Removal of this relict sediment store by dredging starves beaches of new supplies.

In 1897, some 650,000 tons of shingle were dredged offshore of the small village of Hallsands in Devon, England, to provide material for building navy dockyards at Plymouth. The coastal village was protected by an enormous pebble beach, but within a few years of dredging the beach level fell by 12 feet (4 m). With no beach to protect the village from the full force of the sea, storm waves began undermining house and street foundations and destroying houses. After a storm in 1917 villagers were forced to abandon the remains of their 37 homes to the sea.

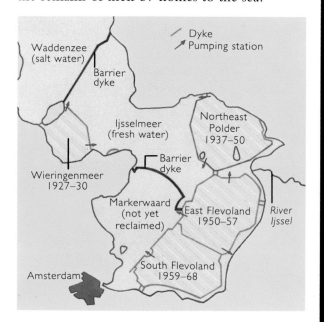

Viewed from space a nation's past misery and current pride is evident as the Netherlands fights to reclaim land stolen by the sea. It is said "God created heaven and Earth but the Dutch made the Netherlands". Although the nation is winning its battle, it has suffered setbacks such as the 1953 North Sea floods which killed 1,835 people and made 100,000 people homeless. More than one-third of the country lies below sea level and one in four people lives below sea level.

In the past 800 years, some 3,000 square miles (7,770 sq km) have been reclaimed. The largest project involved transforming the Zuider Zee, a vast shallow estuary, into a freshwater lake, Ijsselmeer. This was achieved in 1932 by building a 20-mile (32-km) barrier dam. Ijsselmeer is being reclaimed in huge sections, or polders, using ditches, canals and pumps.

The dark green Markerwaard polder can be seen in its early phase, while the speckled red and green Northeast Polder and the twin Flevolands have been farmed and settled for much longer. The Dutch have a long-term mission to reclaim the Waddenzee, tidal flats north of the Ijsselmeer barrier, by constructing barriers to link together the West Frisian islands.

Islands of coral

Beautiful tropical reef islands, with palm-fringed sandy beaches encircled by tumbling white surf, are created by millions of small calcium-secreting organisms called coral polyps. For many thousands of years vast colonies of these marine animals, together with tiny plants called coralline algae, have manufactured limestone from sea water in order to build a reef just below the surface. Small islands then form on this limestone platform from accumulations of sand and debris eroded from the living reef by wave action. Over time, these low-lying islands are colonized by grasses, shrubs, palm trees and mangroves.

The coral animal is a small polyp, related to a sea anemone, ranging in size from $1/10$ inch (2.5 mm) to several inches. It begins life as a free-floating larva but settles in one place to reach its adult stage and then constructs a limestone coral tube in which to live. Reef-building coral polyps live in large colonies and when one polyp dies, its skeleton provides the foundation for a new one to grow on top of it. In this way, the reef grows upward to the low tide limit.

A coral polyp lives in symbiotic association with brown photosynthesizing algae (zooxanthellae): each organism depends upon the other. The coral provides the algae with a protected environment on the inner surface of the polyp tube; in return, the algae satisfy many of the coral's nutritional needs and replenish oxygen in the water through the process of photosynthesis. Red coralline algae also live on the coral, secreting lime and helping to build up the reef.

Reef-building corals thrive only in warm shallow seas with temperatures in the range 75–84°F (24–29°C). The water must be clear so light can reach the photosynthesizing algae. Most coral growth takes place within 65 feet (20 m) of the surface and seldom extends more than 150 feet (45 m).

Nearly 3,000 individual coral reefs form the Great Barrier Reef, the world's largest living structure, which runs parallel to the Australian continental shelf and extends for 1,250 miles (2,010 km). In terms of the number of species, the reef supports the richest ecosystem on the planet including one third of all fish species.

Coral reefs are under attack from growing numbers of crown-of-thorns starfish which devour the coral polyps. These animals have damaged one third of the Great Barrier Reef. Increased inshore amounts of nutrients derived from agricultural fertilizers and sewage appear to increase the survival of the starfish larvae. Mining the limestone reefs for building materials and the effects of tourism and commercial fishing are also contributing to the destruction of the coral reefs.

The Ihuru Islands are part of the chain of nearly 2,000 coral reefs and atolls which form the Maldives in the Indian Ocean. These islands, none rising to more than 6 feet (1.8 m), are home to 250,000 people.
Increasing concentrations of greenhouse gases in the atmosphere are causing the oceans to warm. As a result the sea water is expanding and the world sea level is rising, threatening to drown the islands.

In 1837, Charles Darwin suggested that a coral reef forms initially around a volcanic island. As coral growth keeps pace with the subsiding sea floor and volcanic island or with the rising sea level, a fringing reef is transformed into a barrier reef with a lagoon between it and the island. Eventually the sea floor subsides (or the sea level rises) and the island disappears, leaving only a ring-shaped reef or atoll encircling a lagoon. This finally subsides, forming a guyot below the surface.

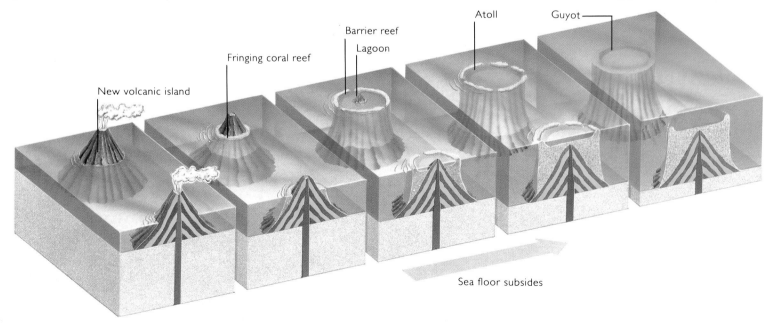

New volcanic island
Fringing coral reef
Barrier reef
Lagoon
Atoll
Guyot
Sea floor subsides

The atmospheric envelope

Earth's atmosphere is essential for our survival because it not only provides us with life-giving oxygen but also protects us from the harmful radiation and galactic debris which bombard our planet. The atmosphere provides a mobile medium for the circulation of vital nutrients and water, and maintains the relatively narrow range of surface temperatures within which the human species can thrive.

The gases which form the atmospheric envelope include nitrogen (78.08 percent by volume), oxygen (20.95), argon (0.93) and carbon dioxide (0.036). Water vapour occurs in variable amounts and trace gases include neon, hydrogen, helium, ozone, methane and nitrous oxide. Several gases, including water vapour and carbon dioxide, are called greenhouse gases because they trap a portion of the Earth's outgoing infrared radiation and radiate it back to the surface, keeping the surface warmer than it would be otherwise.

The atmosphere is not one amorphous mass. Rather it consists of several layers, each with distinct characteristics. The layer closest to the surface is the troposphere, which contains most atmospheric moisture. The next layer, the stratosphere, is dry, and contains large amounts of ozone gas. Ozone protects life on the surface by absorbing most of the Sun's harmful ultraviolet rays.

Above the stratosphere lies the cold and inhospitable mesosphere. Although it contains percentages of nitrogen and oxygen similar to those at sea level, the density of gases is so low that we would not survive long. Each breath would contain far fewer oxygen molecules, so we would suffocate, and exposure to the Sun's ultraviolet rays would cause severe burns.

The top of the mesosphere is the coldest region in the atmosphere (−130°F/−90°C) and is marked sometimes by noctilucent clouds. These silver, rippling veils of cloud appear in the dark sky after sunset, illuminated by the Sun over the horizon.

Above 50 miles (80 km) lies the intensely hot thermosphere. This is followed at 300 miles (480 km) by the rarefied exosphere, which then merges into outer space as air molecules become increasingly scarce.

The troposphere is the lowest and thinnest atmospheric layer and contains virtually all atmospheric moisture and consequently all the clouds, rain, snow and weather.

Sometimes, tiny amounts of moisture escape into the next layer, the dry stratosphere. This results in the formation of cirruslike clouds at altitudes of 12–20 miles (20–30 km) called nacreous, or mother-of-pearl, clouds because of their seashell-like iridescence.

Information on air pressure, temperature and moisture conditions within these two layers comes from a radiosonde instrument. This device is attached to a gas-filled balloon and released daily. As the balloon rises, a radio transmits vital weather information to the surface which meteorologists use to prepare weather forecasts.

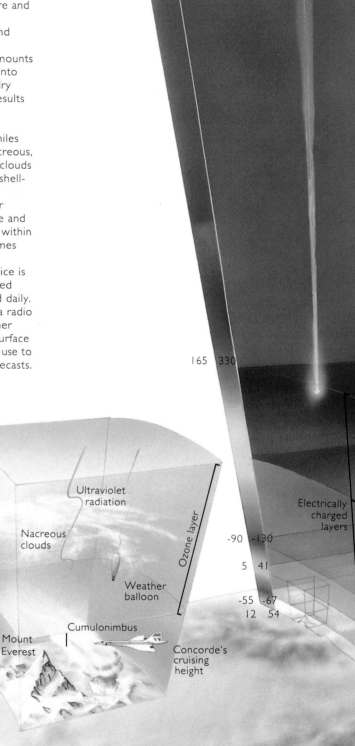

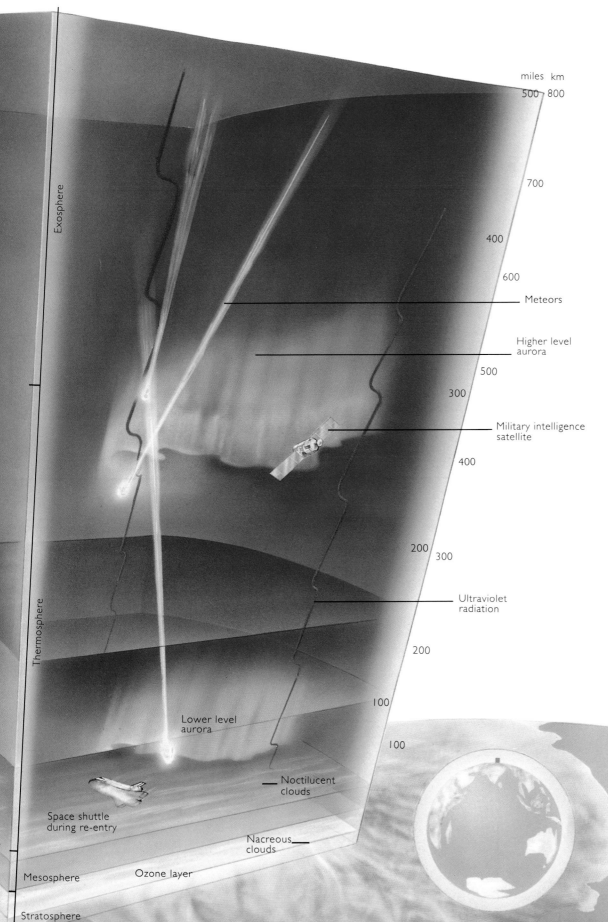

miles km

500 — 800

700

400

600

—— Meteors

Higher level
aurora

500

300

Military intelligence
satellite

400

200
300

Ultraviolet
radiation

200

100

Lower level
aurora

100

—— Noctilucent
clouds

Space shuttle
during re-entry

Nacreous
clouds

Ozone layer

Exosphere

Thermosphere

Mesosphere

Stratosphere

Troposphere

The envelope of gases
forming the atmosphere
appears remarkably thin
compared with the size of
the planet, given that it
plays such a vital role in
sustaining life on Earth.

Although the
atmosphere extends 1,500
miles (2,400 km) above
the surface, it is the 10
miles (16 km) closest to
the Earth that contain 75
percent of the total mass
and virtually all the
moisture. This is the
troposphere.

Superimposed on the
mesosphere and
thermosphere is an
electrified layer, the
ionosphere. Incoming
radiation knocks electrons
off atoms, producing a
mixture of positively
charged atoms (ions) and
free electrons which are
concentrated into bands.

These electrically
charged bands create a
faint airglow at night and
are used in radio
broadcasts. They reflect
radio waves to distant
locations which would
otherwise not receive the
signals since radio waves
travel in straight lines
whereas the Earth is a
sphere.

The ionosphere is also
lit by colourful shimmering
auroras, triggered by solar
flares which send a rush
of electrically charged
particles toward Earth.
The particles cluster
around the magnetic poles
where they collide with
molecules and atoms in
the ionosphere, exciting
them so they release
energy in the form of
different coloured lights.

139

Cloudscapes

Clouds are the visible signs of the atmospheric heat engine at work. They reveal, for example, hot air rising as the Sun heats the ground surface. The shape and depth of clouds also give clues to the temperature and moisture content of the air. Because they bring the rains needed for life to survive, clouds are an essential part of the planet's water cycle. And the movements of clouds indicate the direction and speed of winds as they transfer warm air poleward and cold air equatorward as part of a global energy transfer network.

Clouds are composed of countless millions of minute water droplets or ice particles suspended in the atmosphere. They arise when moisture in the air, in the form of water vapour, approaches its saturation point. Provided suitable seeding nuclei, such as sea-salt or soil particles, are present in the air, the water vapour will be deposited around a particle, forming a minute water droplet, in a process called condensation. If temperature is well below freezing, the water vapour may form directly as ice around a nucleus, creating an ice particle, in a process called sublimation.

Many processes cause air to approach the point of moisture saturation. Heating of the surface causes air bubbles to rise, just like a hot-air balloon. Since air pressure decreases with height, this rising bubble expands. Expansion causes it to cool and eventually the air will reach moisture saturation at its dew-point temperature. Water droplets form at this height in the atmosphere creating the base above which the cloud develops.

Such expansional cooling of air, so that its saturation point is reached, can occur when air is forced to rise over a hill. The higher the hill, the more likely stratus or hill fog will envelop the peak. A similar cooling process happens when warm air is forced to rise over a wedge of cold dense air, for instance along the warm front of a mid-latitude storm. This results in a broad multilayered band of cloud. Conversely, a narrower band of tall clouds, including cumulonimbus, forms along the cold front of the same storm as warm moist air is scooped upward by the advancing wedge of cold air.

Clouds can frequently be seen suspended above hot spots such as power stations, burning oil wells, active volcanoes and blazing forest fires.

How clouds form

Small cumulus cloud forms

Cloud detaches from rising bubble and moves away

Wind

Condensation level

Wind

Wind

Bubble of warm air rises over bare land

Thermal

New bubble forms

On a sunny day, warm air in contact with the ground rises from the heated surface, generating fast-rising bubbles of air. Some surfaces warm faster than others, including dark soils and bare rock.

Air pressure decreases with height, so a rising bubble encounters lower pressure which expands its volume, making it cool down. As it does so, the air approaches its saturation point which means that its gaseous moisture condenses to form water droplets.

The height at which this occurs is called the condensation level and marks the altitude at which a fair-weather cumulus begins to develop. Eventually the wind detaches the cloud from its rising bubble and it drifts away.

If a succession of rising bubbles enters the cloud, it can grow both vertically and laterally. When the supply of bubbles ceases during the evening, the clouds begin to dissipate.

Cloud types

The shape of clouds and the altitude at which they form provide the means of recognizing the 10 basic types of cloud.

Cumulus are small fluffy-white clouds which form on warm sunny days. They show limited vertical growth but can be transformed into a turbulent **cumulonimbus** extending the full depth of the weather atmosphere (troposphere). Jet stream winds stretch its crown into an anvil shape. It brings stormy showers of rain, hail and snow together with thunder and lightning.

Stratus clouds are low level, uniform grey clouds often enveloping hills and producing drizzle. **Stratocumulus** appear as low lumpy-looking layered clouds with bands of dark and light shading. **Nimbostratus** are gloomy, rain-bearing clouds which hug the ground.

Altocumulus, medium-height clouds, are made up of islands of cloud segments creating a mackerel sky, so-called because it resembles the markings on this fish's back.

Altostratus are middle-level greyish clouds which obscure the Sun slightly, making it appear watery.

Cirrus, found at the very top of the weather atmosphere, are the wispy white clouds blown by the wind into delicate banners, hooks and mare's tails. **Cirrocumulus** are high-level ice crystal clouds made up of small rounded white puffs, creating a dappled or rippled appearance.

Cirrostratus are widespread thin clouds formed high in the sky. Their ice crystals scatter light and create a halo round the Sun or Moon — often the only clue to the clouds' presence.

Cloudscapes/2

These generate vast amounts of moisture and seeding nuclei, as well as strong upcurrents.

Under clear skies, night-time cooling of air at the surface can chill the air to its dew-point temperature, causing water droplets to form. The resulting fog or mist is simply a cloud hugging the ground. When night-time cooling occurs in a valley, cold dense air flows down the slopes creating dense fog in the valley bottom. Smoke particles from chimneys are excellent seeding nuclei and the fog (or smog) can become very dense and may present driving and health hazards.

Clouds take many forms but are grouped into ten main types according to their shape and the general height at which they form. Some clouds can display extraordinary shapes. Cumulonimbus are storm clouds, often with a distinctive anvil-like top, caused when the cloud reaches the atmosphere's jet stream and

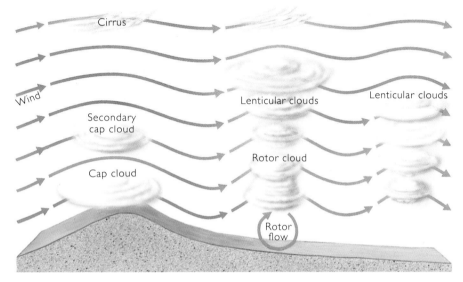

Moist air forced to rise over a mountain range condenses to produce a cap cloud enveloping the peaks. The air flowing down the leeward slopes is compressed as it meets higher air pressure at lower altitudes, making it warmer and drier, and so clouds do not form.

The rising and sinking of air over a mountain may start a wavelike pattern downwind, with a localized reversed flow, or rotor flow. As air rises toward the crest of a wave it cools and condenses, creating small lenticular clouds. Where the air sinks into a trough the cloud dissipates.

When there are distinct layers of dry and moist air flowing over the mountains, several tiers of cloud occur producing, for example, secondary cap clouds and, at higher levels, thin cirrus cloud.

Cold dense air flows off Greenland, across the sea ice. As it passes over the warmer sea, shallow clouds form, which align themselves with the wind into "cloud streets" (left).

Near the limit of the sea ice lies the isolated island of Jan Mayen with a peak of 7,471 feet (2,277 m). When the air flow reaches Jan Mayen buoyancy is not sufficient to lift it over the island, and it is forced to flow around it, creating swirling eddies or vortices.

is pulled out or flattened by the wind current.

Thin lens-shaped, or lenticular, clouds may be seen near mountains. The clouds remain in the same position even though air is rushing through them. This cloud type forms in the crests of waves that develop downwind of a mountain. On rare occasions, several of these clouds can be piled on top of one another resembling a stack of pancakes or, as many people claim, a UFO.

At high altitudes, aircraft often produce trailing lines of cirrus called contrails. These clouds form when water vapour condenses from the hot engine exhausts. Contrails are also created when the aircraft wing tips create turbulent eddies within which the reduced air pressure causes expansional cooling and cloud formation. These tubes of cloud can persist for several hours and spread out to form a thin layer of cirrostratus.

Lenticular clouds above Mount McKinley in Alaska. At 20,320 feet (6,200 m) above sea level, this mountain is already awe inspiring; the cloudscape enhances its drama.

Frozen jewels

Delicate ice prisms, beautiful symmetrical snowflakes, large rain droplets and giant destructive hailstones are among the many different sorts of precipitation that clouds can make. It is the temperature and moisture content of the cloud as well as the strength of its upcurrents that determine which type of precipitation falls to Earth.

Clouds are composed of vast numbers of minute water droplets and ice crystals suspended in the air. The individual droplets and crystals are too small and too light to reach the Earth's surface; several million must combine before precipitation falls. Billowing clouds such as cumulus produce precipitation because the frenzy of air within them makes cloud droplets collide and coalesce into larger droplets until they are heavy enough to fall as raindrops. Most rain that falls in the tropics is formed in this way.

In deeper or higher clouds, which have temperatures well below freezing, many water droplets are "supercooled" but unfrozen. When a few tiny ice crystals form among these droplets they cause the surrounding droplets to evaporate and their water vapour freezes on to the ice crystals.

The crystals grow quickly and some splinter to form small crystals which also steal the water vapour from droplets. Many crystals join together to create snowflakes which are heavy enough to escape from the cloud. Some crystals scoop up tiny cloud droplets which freeze on contact, thickening the ice crystal and creating a snow pellet, or graupel.

The crystal lattice of ice ensures that a snow crystal will be hexagonal, but whether it will be a thin plate, a long needle, a tall prism or a six-sided star depends on the temperature and moisture content of the cloud. Cold dry conditions produce needles and prisms, whereas moist warm conditions create delicate snowflake stars.

When heavy snowfall is driven by strong winds or gales, the resulting blizzard may block roads with severe snow drifts, topple trees or damage roofs. On steep mountain slopes lies another snow threat – the avalanche.

Snow crystals can be feathery six-rayed stars or flat hexagonal plates, depending on the temperature and moisture content of the cloud (right). They are so delicate they collide and break up as they fall to the ground, usually arriving as tangled clumps of many crystals.

In warm clouds, with temperatures above freezing point, swirling air currents cause the many millions of tiny cloud droplets to collide and coalesce to produce large drops heavy enough to fall to the ground as rain (left below).

If this rain falls on to a frozen surface it freezes and coats trees, roads and buildings with a layer of clear ice, or glaze. The weight can break tree branches and power lines.

Some shallow clouds fail to produce large raindrops but smaller droplets may reach the ground as drizzle if the cloud base is near the ground.

In deep clouds which extend high into the cold upper atmosphere a few ice particles form among the many tiny water droplets. Once ice particles exist, water vapour shifts rapidly from the droplets on to the ice flakes causing them to grow larger and the droplets to shrink.

In only a few minutes a large ice crystal is created which joins with other ones to form a branching snowflake. The flake falls as crisp dry snow on a cold ground surface, or as wet snow on a warm one. If it passes through a layer of warm air beneath the cloud, it melts to form rain. Most rainfall in middle latitudes starts off in clouds as snowflakes.

How precipitation forms

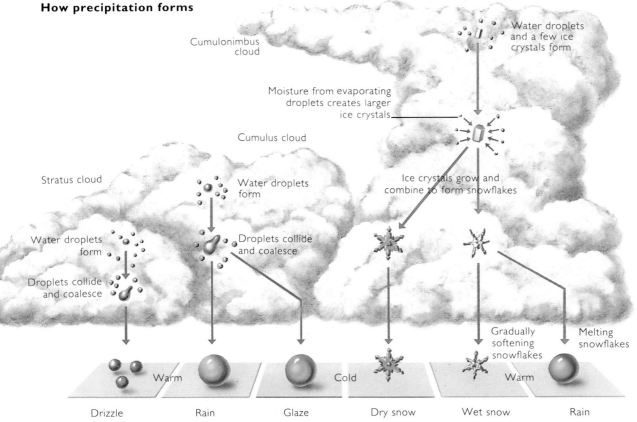

Cumulonimbus cloud

Water droplets and a few ice crystals form

Moisture from evaporating droplets creates larger ice crystals

Cumulus cloud

Ice crystals grow and combine to form snowflakes

Stratus cloud

Water droplets form

Water droplets form

Droplets collide and coalesce

Droplets collide and coalesce

Gradually softening snowflakes

Melting snowflakes

Warm

Cold

Warm

Drizzle   Rain   Glaze   Dry snow   Wet snow   Rain

Frozen jewels/2

When a deep layer of compacted snow overlies a loose layer, vibrations triggered by a skier or a sudden rise in temperature cause the upper layer of snow to surge downslope, sweeping away trees and homes in seconds. Vulnerable towns in mountain valleys may use explosives to set off small avalanches so that a snowfield is unable to create a very large and potentially life threatening avalanche.

In thunderstorms the upcurrents are strong and can support the growth of heavy types of precipitation. Raindrops can grow to their maximum size of about 1/3 inch (1 cm) across, after which they become unstable and break apart. The largest form of precipitation is the hailstone, with its layers of ice like the skins of an onion, which can grow to the size of a golf ball or melon. The stronger the thunderstorm upcurrents, the larger the hailstone.

On rare occasions, hailstones have contained unusual objects. At Bennington, Vermont, in June 1950, and at Bournemouth, England, in June 1983, the hail contained lumps of slag and coal that had been lifted into the storm by strong rising air currents. At Dubuque, Iowa, in June 1882, hailstones as large as oranges fell and two stones contained small frogs which were alive when the ice melted.

In the United States, a "white plague" of hail causes considerable damage from Texas through South Dakota from May to June each year. It pulverizes crops, strips the bark off trees, shatters glasshouses, dents the bodywork of cars, and even kills animals and people. In September 1988 hail the size of goose eggs fell in north China injuring 2,000 people. Two years earlier in central China a series of hailstorms killed 100 people and reportedly destroyed 80,000 homes.

Ice crystals and raindrops can create spectacular colour displays in the sky when sunlight strikes them. Sunlight entering a raindrop is bent, reflected and split into its spectrum of colours. A rainbow is created with red at the outside of the bow and violet on the inside. The ice crystals which form high-level cirrus clouds also reflect and refract sunlight, creating rings, arcs and spots of white light in complex geometric patterns.

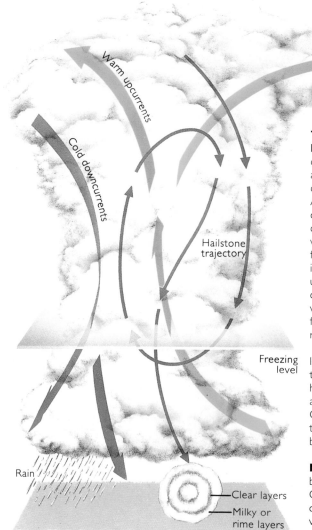

The nucleus of a hailstone is a small ice crystal or one coated with a layer of tiny frozen cloud droplets (graupel). As it falls to warmer parts of the thunderstorm, it collects cloud droplets which freeze slowly, forming a coating of clear ice, or glaze. As it is swept up to colder parts it collides with droplets which freeze on impact, forming a milky layer of rime ice.

The number of ice layers, up to 15–20, indicates the number of times the hailstone was swept up and down within the cloud. Once the hailstones become too heavy to be supported by the storm, they fall.

Massive hailstones bombarded Munich, Germany, in 1984, damaging some 200,000 vehicles, 100,000 homes and 20 passenger jets.

As clouds of freezing fog drift slowly past a cold thin object, some of the tiny water droplets touch the object and freeze immediately. This creates a banner of feathery white ice, known as rime, on the windward side.

Riming is common on the tops of hills and mountains in winter since they remain shrouded in cloud for long periods. Rime banners can reach 10–20 inches (25–50 cm) in width on mountain tops but in the lowlands the banners are seldom wider than about $1/3$ inch (1 cm).

Trees – as shown here in Switzerland – posts, television antennae and telephone wires can gather such heavy ice banners that they bend or snap. Aircraft flying through cloud sometimes collect rime on the wings, which can cause problems if it is very thick and heavy.

Patterns in the air

Winds, clouds and storms are the moving parts of a global atmospheric heat-exchange engine which effectively spreads warmth from the intensely heated equatorial region toward the poles.

Energy from the Sun creates an equatorial furnace which powers the engine and its moving parts, including the trade winds, westerlies, jet streams and cyclones. Because the poles cool rapidly during winter, thereby increasing the pole-equator temperature gradient, the engine operates at a much faster rate in winter than in summer. This explains why winds and storms in middle latitudes are often stronger in winter.

The dominating winds of low latitudes are the persistent trade winds. Although these winds were essential for sailing ships to undertake maritime trading in earlier centuries, their name is actually derived from a nautical term meaning "regular track". Clusters of towering cumulus, thunderstorms and sometimes large swirling storms, or hurricanes, develop within the trade winds.

Beneath the converging hemispheric trade winds lies a narrow band called the doldrums. It is characterized by hot humid conditions, frequent thunderstorms and squally ill-defined winds. The position of the doldrums changes with the seasons as the Sun makes its twice-yearly swing across the equator. Lacking the detailed weather information available to mariners today, early sailing ships often strayed out of the trades and into the nightmare of the doldrums where they drifted aimlessly for days or even weeks on end.

The mid-latitudes are influenced alternately by mild southwesterly winds and cold polar easterlies as the position of the polar front fluctuates both with the seasons and from day-to-day, reflecting the relative strengths of these two opposing winds.

The source area from which winds blow influences their temperature: warm winds come from equatorial regions and cold ones from polar regions. Winds are also affected by the nature of the source area; it may be a continental interior, ocean, snow field, mountainous region or coastal plain.

Three giant circuit loops of winds dominate each hemispheric circulation pattern. The commanding one is the Hadley cell powered by year-round heating of the equatorial regions. Heated air near the surface rises, producing deep clouds and heavy rainfall. As it reaches the tropopause, it is deflected poleward, finally sinking at about 30° latitude.

As the air descends, it encounters greater pressure which squeezes and warms it, wringing it dry of moisture. It is this band of dry sinking air which maintains the arid conditions found in the Atacama, Australian, Kalahari and Sahara deserts.

Some of the sinking air is drawn back toward the equator by low surface air pressure created by the rising equatorial air currents. These returning winds are the northeast (northern hemisphere) and southeast (southern hemisphere) trade winds.

In the Ferrel cell some of the air descending at 30° latitude moves poleward to form the westerlies of the middle latitudes. These winds do not head poleward along a direct south-north alignment because the planet's rotation bends winds to the right in the northern hemisphere and to the left in the southern hemisphere.

As these warm moist winds move into higher latitudes they clash with cold dry winds pushed out from the polar dome of cold sinking air – the polar cell. The fierce battle between these two opposing winds leads to the creation of mid-latitude cyclones (frontal depressions).

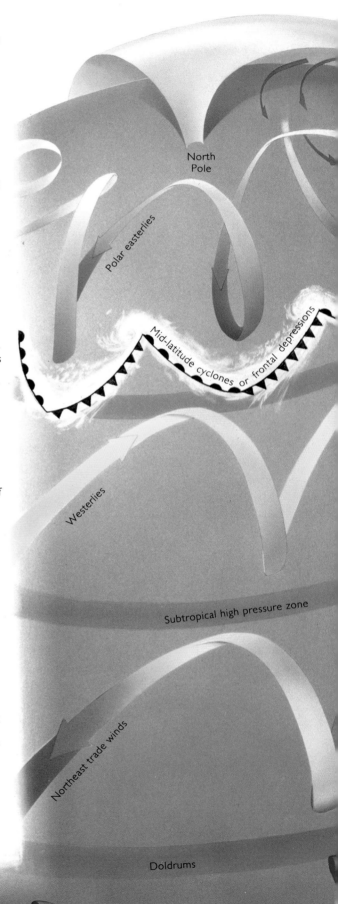

North Pole

Polar easterlies

Mid-latitude cyclones or frontal depressions

Westerlies

Subtropical high pressure zone

Northeast trade winds

Doldrums

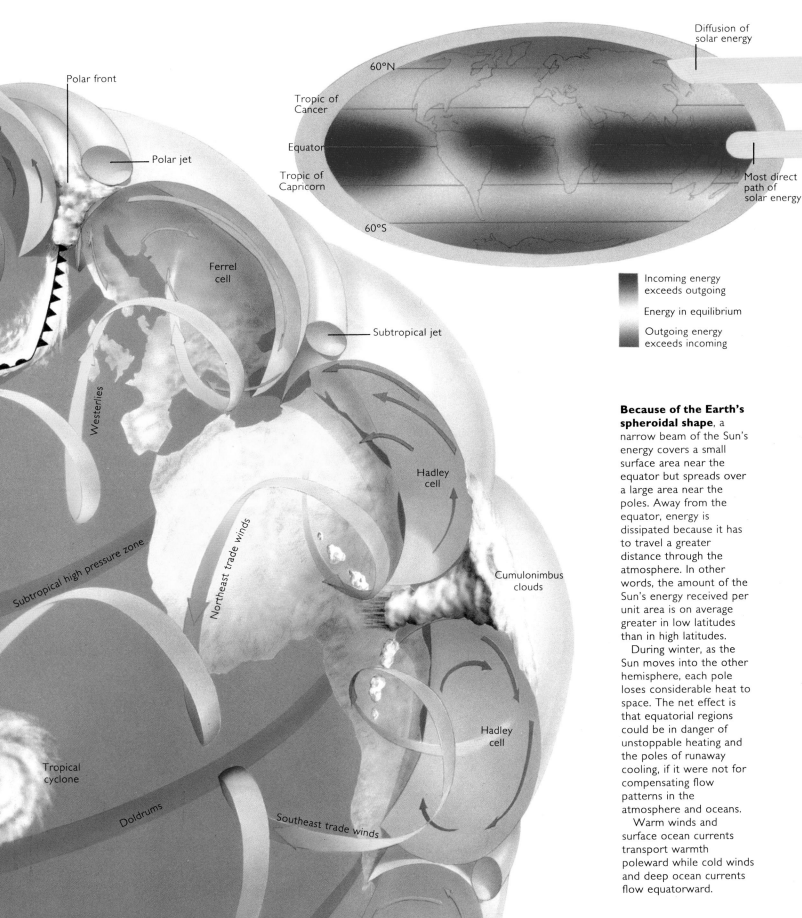

Diffusion of solar energy

60°N

Tropic of Cancer

Equator

Tropic of Capricorn

Most direct path of solar energy

60°S

Polar front

Polar jet

Ferrel cell

Subtropical jet

Westerlies

Subtropical high pressure zone

Northeast trade winds

Hadley cell

Cumulonimbus clouds

Hadley cell

Tropical cyclone

Doldrums

Southeast trade winds

Ferrel cell

Incoming energy exceeds outgoing

Energy in equilibrium

Outgoing energy exceeds incoming

Because of the Earth's spheroidal shape, a narrow beam of the Sun's energy covers a small surface area near the equator but spreads over a large area near the poles. Away from the equator, energy is dissipated because it has to travel a greater distance through the atmosphere. In other words, the amount of the Sun's energy received per unit area is on average greater in low latitudes than in high latitudes.

During winter, as the Sun moves into the other hemisphere, each pole loses considerable heat to space. The net effect is that equatorial regions could be in danger of unstoppable heating and the poles of runaway cooling, if it were not for compensating flow patterns in the atmosphere and oceans.

Warm winds and surface ocean currents transport warmth poleward while cold winds and deep ocean currents flow equatorward.

149

Patterns in the air/2

As winds flow out from these areas they subject other areas to their air mass characteristics in the sense that "every wind has its weather".

The air mass affecting an area can change through the seasons with, for example, a tropical maritime air mass dominating in summer and a polar continental air mass in winter. Because the global pattern of winds and air masses remains similar from year to year, every part of the world experiences a particular yearly cycle of daily weather through the seasons, termed its climate.

Although atypical extremes of daily temperature and precipitation can occur, there are sufficient similarities from one year to the next to consider that a particular place experiences a distinctive type of climate. Sites around the world with similar climates can be grouped into a limited number of climate zones.

Early climate zone classifications lacked detailed climate information from some regions of the world so vegetation or soils were used as surrogate climate indicators. This explains why some classifications include names such as boreal, steppe and tundra.

Locations regarded as belonging to the same climate zone can be found far apart. For example Mediterranean or west coast subtropical climate zones, which have warm dry summers and mild wet winters, can be found in Australia, California, Chile and South Africa as well as around the Mediterranean Sea.

Climate zones are among the most important characteristics of the planet, distinguishing areas of different landscapes, agriculture, plant and animal ecosystems and human comfort. Temperate, polar and tropical climate zones each conjure up very different images of landscapes and ways of life.

If the global climate changes, climate zones migrate. When extensive ice sheets covered the continents about 18,000 years ago, the world's climate zones shifted equatorward and were squeezed together. Current global warming due to the greenhouse effect is beginning to shift climate zones poleward.

These changes may be taking place so rapidly that some plants and animals, having evolved within a specific climate zone, may be unable to adapt and risk extinction.

Surface winds over the Pacific Ocean derived by satellite measurements of radar reflections from small wind-driven waves on September 14, 1978.

Arrows point in the direction the winds blow. Areas of light winds are coloured blue and strong winds orange. In mid-Pacific the strong trade winds meet along a snaking boundary denoting the doldrums. Tight wind spirals indicate intense storms over the Southern Ocean while an area of light winds in the North Pacific marks a large high pressure anticyclone.

The global pattern of winds and the influence of mountains, continents and ocean currents produce a limited number of distinctive world climate zones.

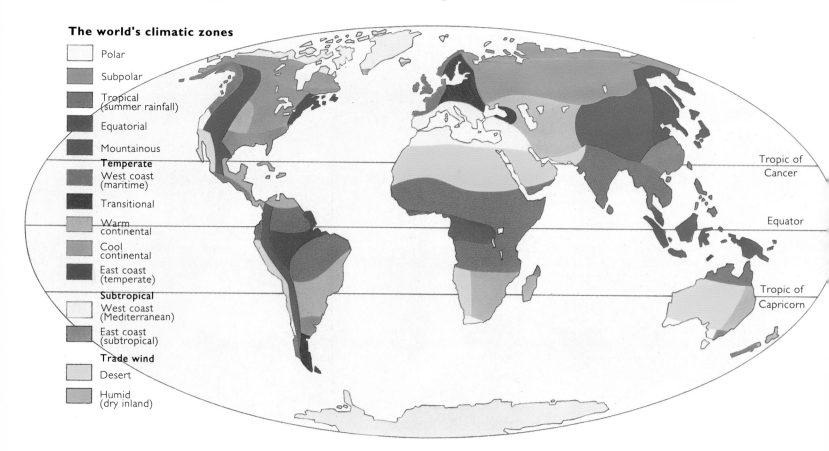

The world's climatic zones

- Polar
- Subpolar
- Tropical (summer rainfall)
- Equatorial
- Mountainous

Temperate
- West coast (maritime)
- Transitional
- Warm continental
- Cool continental
- East coast (temperate)

Subtropical
- West coast (Mediterranean)
- East coast (subtropical)

Trade wind
- Desert
- Humid (dry inland)

Tropic of Cancer

Equator

Tropic of Capricorn

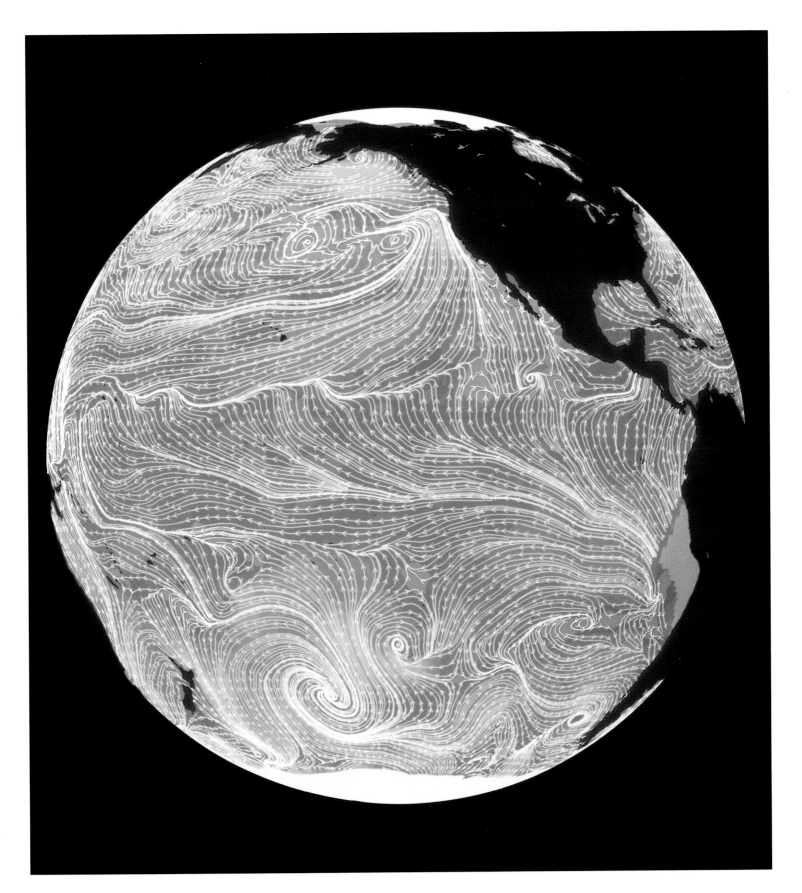

Monsoons

Three billion people – half the world's population – depend for their survival on the seasonal reversal of winds, known as the monsoonal circulation. For six months of each year, the countries of Southeast Asia are exposed to relentless drought. Toward the end of this dry season, food shortages worsen, water supplies dwindle and power produced by hydroelectricity generators declines.

Then suddenly torrential rains burst forth upon the parched landscape with the start of six months of the monsoonal wet season. Rural communities come alive as farmers leap into action and water and power supplies are replenished. The wet season marks the renewal of life and the promise of prosperity.

The cause of the monsoonal circulation lies in the unequal heating of the Earth's tropical continents and oceans as the overhead Sun swings back and forth in its annual cycle between the two hemispheres. The Southeast Asian monsoon is the largest and strongest monsoonal circulation. In India, the rains arrive first at its southern tip on June 1. Like an expectant parent, the nation awaits news of its arrival and confirmation that the worries and troubles of the dry season will soon be over.

By September the squally rains have progressed northward to vent their fury against the Himalayan foothills. With so much rainfall, rivers are transformed into raging torrents. The mighty Ganges and Brahmaputra rivers meet in low-lying Bangladesh and extensive floods occur there each year. In September 1988 two-thirds of the country was inundated, making 28 million people homeless.

Heavy monsoon rains in the eastern Indian state of Orissa cause extensive floods, but without these rains agriculture would fail and famine follow.

The rains arrive in Orissa on June 10 from the same storms which then move north to the head of the Bay of Bengal and dump vast amounts of rain on the Khasi Hills.

Cherrapunji is here, with its world record 12-month rainfall total of 1,041.78 inches (26,461 mm) between August 1860 and July 1861.

During the northern hemisphere winter the Sun is overhead in the southern hemisphere, and the northern continents cool while the southern oceans and lands warm (below).

Cool dry northeasterly winds flow outward from the Asian and north African interiors, creating a cool dry season there. As these trade winds cross the equator they are deflected east by the Earth's rotation. As they flow over the oceans they gather moisture and deliver torrential rains, creating the wet season for Indonesia, Borneo, Sumatra and northern Australia, as well as west central Africa and central South America.

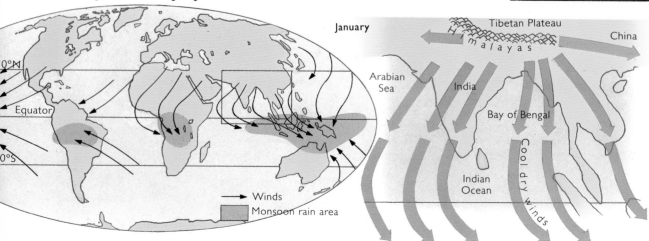

January

Winds

Monsoon rain area

Tibetan Plateau

China

Himalayas

Arabian Sea

India

Bay of Bengal

Indian Ocean

Cool dry winds

The monsoonal circulation reverses in the northern hemisphere summer. This is most marked in Southeast Asia where the land, especially the Himalayas and Tibetan Plateau, warms rapidly compared with the oceans. Warm air rises over the land, drawing in moist air in the form of a sea breeze. From May to September, southwesterly winds bring heavy rains to most of Southeast Asia.

152

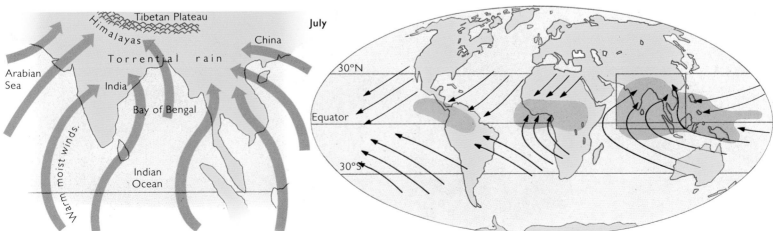

July

Tibetan Plateau

Himalayas

China

Torrential rain

Arabian
Sea

India

Bay of Bengal

Warm moist winds.

Indian
Ocean

30°N

Equator

30°S

Jet streams: fast lanes in the sky

B ands of cirrus clouds, about 200 miles (320 km) wide, race high across the sky in middle latitudes indicating the presence of swift westerly winds circling the Earth. These fast-flowing rivers of air follow sinuous paths, meandering first poleward and then equatorward. They are a vital part of the global atmospheric heat-exchange engine, transferring warm tropical air toward the poles and cold polar air toward the equator.

The most important jet stream forms at the mid-latitude boundary, the polar front, which divides cold polar air from warm tropical air. Surface air pressure is slightly lower on the cold side of the front than on the warm side. Since air pressure decreases with height more quickly in cold air than in warm air, the small pressure difference at the surface becomes an enormous difference across the front in the upper atmosphere.

The horizontal pressure gradient at a height of about 6-9 miles (10-15 km) generates a narrow zone of exceptionally strong winds. This polar front jet, deflected eastward by the Earth's rotation, is only 180-300 miles (290-480 km) wide and 1-2 miles (2-3 km) deep. It often reaches speeds of 100-150 mph (160-240 km/h); in winter, however, when the temperature contrast across the polar front is greatest, speeds may exceed 300 mph (480 km/h).

The jet stream follows a wavy path around the Earth. In winter when the atmospheric

Photographed from a Gemini spacecraft, long ribbons of cirrus clouds (below) are seen streaking eastward across the sky over Egypt and the Red Sea at speeds exceeding 100 mph (160 km/h). Such clouds reveal the presence of an invisible jet stream, a tubelike core of very fast winds, which encircles the planet. This one is the subtropical jet stream found at a height of 8 miles (13 km) and is located at the poleward edge of the Hadley cell.

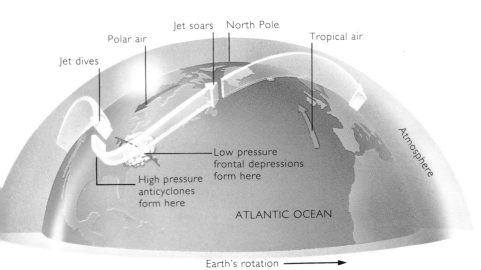

Jet dives
Polar air
Jet soars   North Pole
Tropical air
Atmosphere
Low pressure
frontal depressions
form here
High pressure
anticyclones
form here
ATLANTIC OCEAN
Earth's rotation

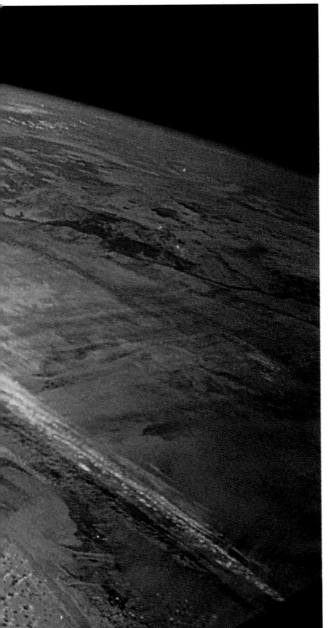

The polar jet stream meanders from side to side as it races round the Earth (above). Lobes of warm tropical air push poleward producing ridges in the jet stream, while cold polar air bulges equatorward forming troughs.

Where the jet flows equatorward into a trough it dives, narrows and presses down on the underlying air. This leads to high air pressure at the surface. Air flows outward from this region, swirling clockwise, forming an anticyclone with clear dry weather.

Where the jet stream flows poleward out of a trough it soars in altitude, broadens and sucks air upward. This creates low pressure at the surface pulling air toward it in a counterclockwise direction. The rising air leads to cloud formation and the first stage of a mid-latitude frontal depression, or cyclone. The faster the jet stream the more intense the storm. Once formed, the storm is steered northeastward by the jet stream.

heat-exchange engine is most vigorous, the strengthened jet shifts a few degrees toward the equator and displays few waves or zigzags. In summer, the weaker jet shifts poleward and exhibits as many as six waves.

Where the polar front jet bends poleward in its winding path it triggers an endless succession of swirling mid-latitude storms which sweep across the continents and oceans guided by the jet stream. It is these frontal depressions, or cyclones, that bring the changeable weather conditions to temperate latitudes. Weather forecasters rely on locating the jet stream to see where these storms form and the track they will take.

The polar front jet behaves like a meandering river, and over several weeks its meander loops become accentuated. Eventually, just like a river forming an oxbow lake, the jet stream cuts across the narrow neck of a meander, leaving a pool of warm air in high latitudes or cold air in tropical latitudes. These isolated air masses remain stationary for weeks and are called "blocks" because they divert approaching storms farther north or south than usual, bringing unseasonal weather to those latitudes.

Jet streams were first discovered in November 1944 when 111 high-flying B-29 bombers were sent north from Saipan to bomb industrial sites near Tokyo. As the planes turned to approach the city at a height of 6 miles (10 km), they were swept forward by westerly winds reaching 150 mph (240 km/h), which carried nearly all their bombs out to sea.

Today airlines are fully aware of jet streams. Flight paths across the Atlantic Ocean are planned to take advantage of strong tail winds when travelling to Europe and to avoid powerful head winds when flying to North America. Flights to Europe can save an hour or two using the jet stream.

Jet streams are also found at 30° latitude at about 8 miles (13 km) in each hemisphere. The subtropical jet separates warm tropical air of the Hadley cell from the cooler air of mid-latitudes. The westward-flowing polar and subtropical jets sometimes merge to form one powerful jet stream, for example over China and Japan.

Explosive depressions

Seen from outer space, the middle latitudes of the planet are dominated by an eastward-moving procession of huge comma-shaped storms. Certain regions spawn a succession of these storms which expand, intensify and eventually decay as they sweep eastward during a life cycle lasting several days.

Many of these storms, in different stages of development, link together along a long winding track stretching across the oceans and continents. The storms reach over 1,000 miles (1,600 km) across and are called mid-latitude cyclones, or frontal depressions. They produce the changeable weather so characteristic of temperate climates.

The middle latitudes contain the all-important air mass boundary, the polar front, which separates cold polar air from warm tropical air. These adjacent air masses resist mixing, yet exchange of heat across this boundary is necessary to maintain a global atmospheric heat balance. The key to achieving heat exchange is the frontal depression, or cyclone, because it envelops a portion of cold and warm air along the polar front and, as the storm swirls through its various stages of development, forces the air masses to mix. Each year the collective action of these storms transfers the large amounts of energy needed to maintain the heat balance between the poles and the equator.

Frontal depressions form beneath the polar front jet where the jet stops diving toward the equator and begins soaring poleward. Such troughs along the wavy jet stream are often located in the lee of the Rocky Mountains and off the east coasts of North America and Asia.

The storms born over the Atlantic Ocean south of Newfoundland are the source of much of Europe's weather. Warmer than average water in this area in winter can mean Europe will experience a constant stream of

As a mid-latitude frontal depression develops, the overlying jet stream draws air upward along part of the polar front – the boundary between cold polar and warm tropical air. This lowers the surface air pressure and triggers the wave stage as air is sucked inward, spiralling counterclockwise (1).

The open stage occurs when two segments of the polar front – the warm and cold fronts – rotate around the deepening low pressure centre (2). The cross-section (bottom) shows how warm air rides up over the cold dense air along the warm front, forming clouds.

Clouds progress from cirrus, through cirrostratus and altostratus, to nimbostratus clouds with moderate rainfall. The warm air wedge between the fronts contains only thin clouds before the cold front approaches, bringing cumulonimbus and heavy rainfall.

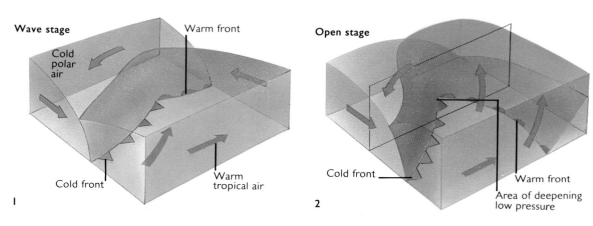

Wave stage — Cold polar air — Warm front — Cold front — Warm tropical air — 1

Open stage — Cold front — Warm front — Area of deepening low pressure — 2

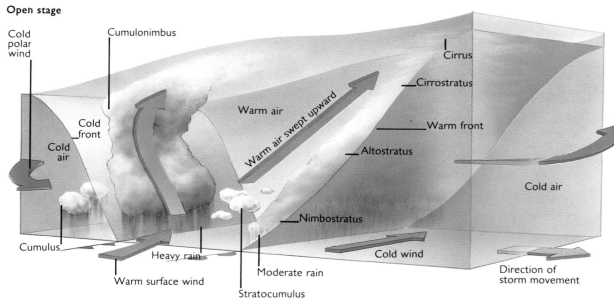

Open stage — Cold polar wind — Cumulonimbus — Cirrus — Cirrostratus — Cold front — Warm air — Warm air swept upward — Warm front — Cold air — Altostratus — Nimbostratus — Cold air — Cumulus — Heavy rain — Moderate rain — Cold wind — Warm surface wind — Stratocumulus — Direction of storm movement

storms bringing mild windy weather, whereas cold water results in fewer storms and the likelihood that Europe will suffer a dry bitterly cold winter.

Plotting the position and strength of the meandering jet stream allows forecasters to predict where storms will be born and how intense they will be. As the storms develop they are guided northeast by the jet stream until at the poleward end of a jet stream wave, where the winds turn equatorward and so clockwise, they begin to slow down and die.

If a storm intensifies rapidly it is called an explosive cyclone. In October 1987 Hurricane Floyd off Florida sent warm tropical air high up into the atmosphere to confront the cold polar jet sweeping across the Atlantic. The temperature contrast accelerated jet stream winds to 200 mph (320 km/h), creating the worst storm to strike England and France since 1703.

A sudden surge of speed in the high-altitude jet stream transformed an ordinary storm into a "bomb" which struck northwest Europe in the early hours of October 16, 1987. This was the worst storm in southern England and northwest France for over 250 years.

Winds reached 94 mph (150 km/h) in London and 120 mph (193 km/h) through the English Channel. A total of 25 people were killed and 120 injured. Forests were flattened and rural landscapes devastated as France and England lost 45 million trees. Property damage, power cuts and economic paralysis cost insurers £1.8 billion.

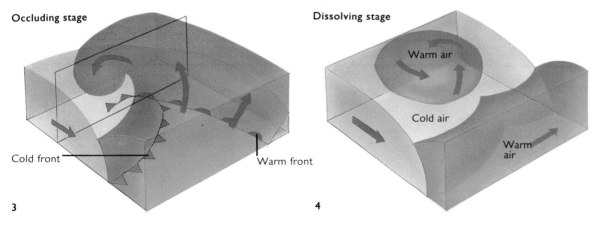

Occluding stage

Cold front

Warm front

3

Dissolving stage

Warm air

Cold air

Warm air

4

The occluding stage begins when the faster-moving cold front begins to catch up with the warm front (3). The wedge of warm surface air is squeezed and pushed aloft. As the cross-section (bottom) shows, the clouds are lifted high above ground and the last of the rainfall is squeezed from these thin clouds.

Eventually all the warm air has been lifted (4). The clouds dissipate, rain stops and the surface fronts of the storm disappear. In this final, or dissolving, stage the polar front re-forms, having shifted slightly equatorward.

The whole cycle takes several days to complete and during this time the storm moves along the polar front at a speed of 25–35 mph (40–56 km/h), guided by the overlying jet stream. Satellite images often reveal a family of frontal depressions lying along the polar front in various stages of development.

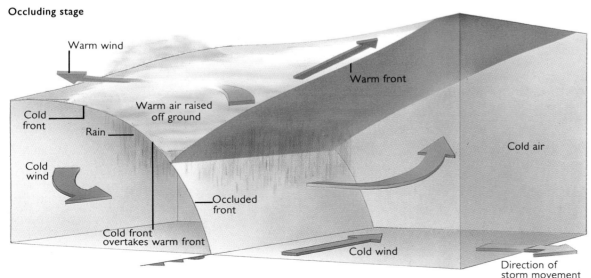

Occluding stage

Warm wind

Warm air raised off ground

Warm front

Cold front

Rain

Cold wind

Occluded front

Cold air

Cold front overtakes warm front

Cold wind

Direction of storm movement

Winds of change

From light breezes to violent hurricane-force gusts, winds blow from areas of higher air pressure to areas of lower pressure. In so doing, they redistribute heat between the equator and the poles, sustain the surface currents of the oceans, guide the clouds and storms which bring rain and snow, drive the waves that assault coastlines, and help sculpt the natural landscape.

Air pressure differences between areas are often the result of temperature differences. Some winds can be localized, such as the light onshore breeze that develops on a sunny day as the land warms up faster than the sea. On the planetary scale the polar easterlies, trades and mid-latitude westerlies exist ultimately because of the temperature difference between the poles and the equator.

Some winds have evocative local names. Hot, dry and dusty winds blowing out from deserts include the sirocco in north Africa, brickfielder in south Australia, and Santa Ana in California. The Santa Ana blows west from the Mojave Desert reaching speeds of 60 mph (96 km/h). In 1982 it fanned fires which destroyed 500 apartment units in Anaheim.

Desert winds sometimes travel long distances before their dusty load is washed out by rain. Iron-stained Saharan sands have occasionally produced "blood rains" as far north as Britain. Funnelling of winds down narrow valleys can create very violent winds such as the mistral in France. Icy blasts sweep down the Rhône Valley in early spring, killing unprotected crops.

Winds can create opportunities as well as threats. More than 300 turbines have been grouped to create a "wind farm" to generate electricity at Altamont Pass, California.

Sea breezes form in coastal locations because land heats up and cools down faster than water.

On a sunny day, land warms quickly and the overlying air becomes buoyant and rises. This creates an area of lower air pressure near the surface which draws in air from the sea as an onshore or sea breeze. It increases in strength in the early afternoon bringing cooler moist air, and sometimes mist.

At night a reversal takes place, with air over the land cooling more quickly than over the water, causing a **land breeze**. Sinking air over the land and rising air over the relatively warmer sea sets up a local circulation. The breeze moving out to sea is sometimes also called an offshore breeze.

As a wind is forced up and over a mountain range it is cooled producing clouds, rain and snow. As this now dry air sinks down the leeward side it is compressed by the higher air pressure of lower altitudes which warms it considerably — sometimes by 36°F (20°C). Warming it makes it drier creating a **rain shadow**.

This warm dry wind is known as the föhn wind in the Alps, the zonda in the Andes foothills and the chinook or "snow eater" east of the Rockies.

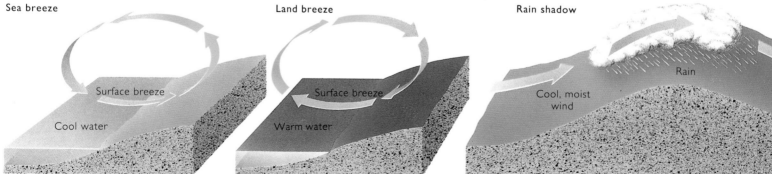

Sea breeze

Surface breeze

Cool water

Warm land

Land breeze

Surface breeze

Warm water

Cool land

Rain shadow

Rain

Cool, moist wind

Sunrise near Townsend, Tennessee, (left) reveals fog-shrouded valleys between the tranquil ranges of the Appalachian Mountains.

Valley breezes are created during the day when air overlying sunny mountain slopes heats up faster than air in the shaded valley below. As it warms it rises, creating lower air pressure than in the valley bottom. Air is then drawn up the slopes and a circulation of air is created within the valley. When this circulation also draws air into the valley an up-valley breeze is generated.

At night mountainside air cools more quickly than air near the valley floor, creating a **mountain breeze.** The chilled air sinks down the valley sides and pools in the valley bottom often producing mist or fog. If a valley floor slopes quite steeply the cold air flows out as a down-valley wind.

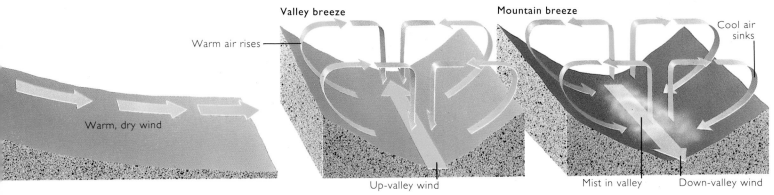

Valley breeze

Mountain breeze

Warm air rises

Cool air sinks

Warm, dry wind

Up-valley wind

Mist in valley

Down-valley wind

159

Thunder and lightning

Flashes of lightning striking the Earth's surface at this very moment from 2,000 thunderstorms scattered across the world are an integral part of a global electrical circuit. Since electrons leak continually from the planet's surface to the air during fair weather, lightning is needed to replenish the Earth's negative field and maintain a balanced electrical state.

While performing its role as a giant electrical discharge generator, the thunderstorm produces deafening roars of thunder, torrential rain, squally bursts of wind, damaging hail and sometimes even violent tornadoes and waterspouts. Some thunderstorms generate strong downcurrents called microbursts which strike the ground with brief but destructive bursts of energy. They flatten crops and trees, wreck buildings and cause aircraft to crash just as they are taking off or landing, as happened at Dallas in 1985 when 134 people were killed.

A thunderstorm develops when strong warm air currents surge upward to create a cumulonimbus cloud several miles across and up to 9 miles (15 km) high. Jet stream winds shred the cloud top into the familiar anvil shape. Thunderstorms can sometimes be lonely giants but at other times several may cluster together or march in line abreast.

Few can imagine what it is like to be tossed about by the violence of a thunderstorm but Lieutenant Colonel William A. Rankin knows first-hand. In 1959 he bailed out of a crippled aircraft over Norfolk, Virginia, into a thunderstorm at an altitude of 9 miles (15 km). A long terrifying free fall took him down to about 2 miles (3 km) where his parachute opened – only 10 minutes from safety in normal circumstances.

He was then blasted upward into the heart of the storm and continued to be swept up and down as if he were a hailstone being formed. For 40 minutes his bruised and shaken body was pelted by freezing rain and cut by hailstones while his senses were assaulted by deafening thunder and flashes of white-hot lightning. Incredibly, he survived.

Lightning from thunderstorms strikes the planet over 100 times every second of the day. Each brilliant flash of lightning is a high-voltage electrical spark carrying tens of thousands of amperes to the Earth – enough energy to light a small city for several weeks. This surging current superheats a column of air 2 inches (5 cm) wide to 54,000°F (30,000°C).

Cloud to air lightning

Electrified "balls of fire" (ball lightning) can appear suddenly inside a building or aircraft during a thunderstorm. These luminous spheres, ranging in size from an orange to a football, are commonly red or orange and last less than a minute.

While floating in the air they appear to move with the air currents, which explains why they seem to be attracted to terrified witnesses.

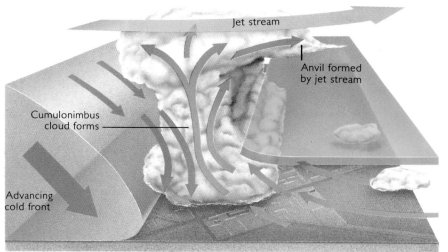

Small cumulus cloud

Thermal lid

Hot spot

Warm moist air

Jet stream

Anvil formed by jet stream

Cumulonimbus cloud forms

Advancing cold front

Thunderstorms result from the Sun's intense heating of sunward-facing hillsides, dark fields and urban areas. As daytime heating of these hot spots intensifies, thermals bubble upward and form small cumulus clouds.

Further cloud growth can be prevented by the presence of an overlying layer of cool dry air (above left), which acts like a lid inhibiting further convection. This thermal lid is often strengthened by subsiding air aloft, yet if upcurrents could penetrate it, their rise would go unchecked and clouds would deepen.

The means to force open the lid can either be very strong solar heating of the ground or, more likely, an advancing cold front (above right). As this dense mass of cold air approaches, it forces air upward which breaks through the lid. A cumulus billows up to form a towering thunderstorm as upcurrents reach speeds of 60 mph (96 km/h). A jet stream aids growth and shreds the top of the cloud into an anvil shape.

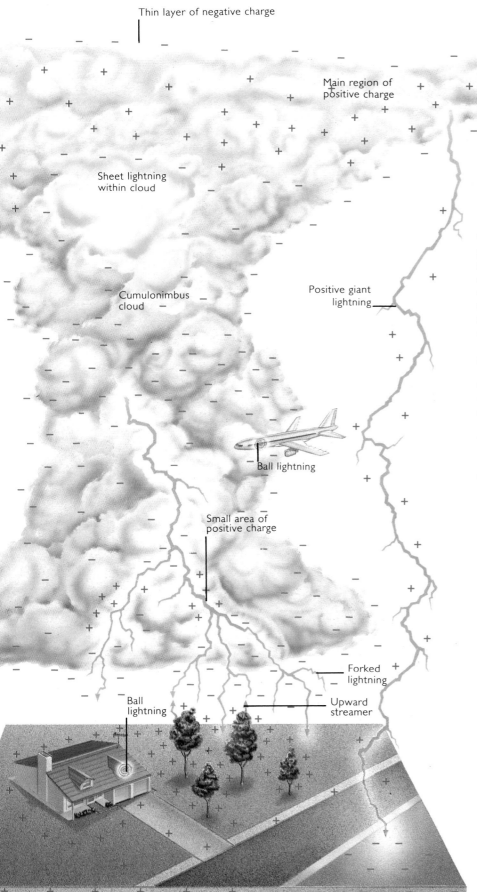

Thin layer of negative charge

Anvil

Main region of positive charge

Sheet lightning within cloud

Cumulonimbus cloud

Positive giant lightning

Ball lightning

Small area of positive charge

Ball lightning

Forked lightning

Upward streamer

Within a thunderstorm, positive and negative electrical charges become separated as positively charged tiny ice crystals and splinters are swept upward while large droplets and ice pellets (graupel) carrying a negative charge descend.

Charges are acquired in various ways. Tiny ice crystals gain a positive charge when they strike large ice pellets. Rapid freezing of a water droplet produces a positive outer shell and, if this fractures, tiny ice splinters carry positive charges aloft while the heavier, negatively charged inner core descends.

Once the electrical charge separation reaches about 100 million volts, lightning is triggered. The commonest form is sheet lightning, which is seen only as a brightening of the sky since it takes place wholly within the cloud. A cloud to ground electrical discharge is the familiar forked lightning. An unusual positively charged giant "superbolt" leaps from the anvil to negatively charged ground ahead of the storm base.

Occasionally, lightning appears out of a clear sky. If a storm is hidden from view by a nearby hill, cloud to air lightning, travelling horizontally a long distance before descending to the ground, can arrive unexpectedly as a "bolt from the blue".

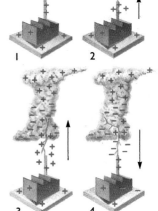

1   2

3   4

A passing thunderstorm induces a positive charge at the ground so a stream of electrons begins searching for a conducting route to the surface (1).

As this faint discharge nears the ground it attracts short streamers of positive charges from tall objects trying to establish a flow of current (2).

When the channels connect, creating a path of least electrical resistance, a powerful bright return stroke streaks upward from the ground in an incandescent flash. (3)

This return stroke is followed by the first negative discharge to the surface. Several return strokes and downward leaders alternate along the channel making lightning appear to flicker (4).

Thunder and lightning/2

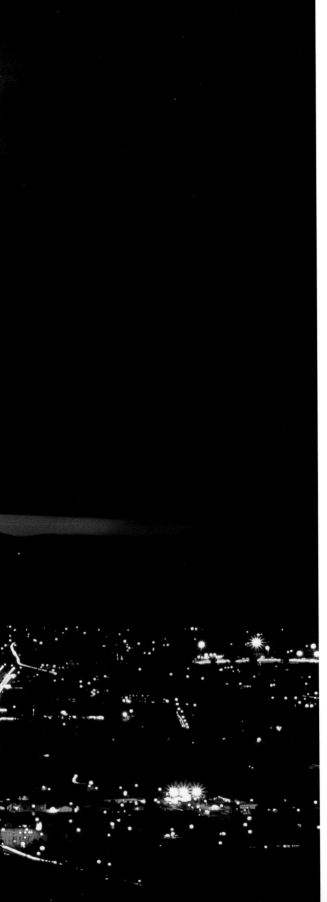

Beautiful but potentially lethal flashes of lightning strike the town of Tamworth, New South Wales, Australia. Some thunderstorms generate as many as 100 lightning flashes a minute.

Lightning comes in many colours. In the United States, where 10,000 lightning-triggered forest and grassland fires occur each year, forest rangers consider white lightning is more likely to start fires than red lightning. Red light is released when hydrogen atoms in raindrops are "excited" by the intense heat of lightning. By contrast, white lightning indicates an absence of rain which could dampen a fire. Icy blue lightning is associated with hail while yellow lightning can indicate a dust-laden sky.

Buildings can be protected by attaching a metal lightning rod to the highest point. This is then connected to a wide copper strip leading down the outside of the building into the ground. Such rods do not suppress lightning but they provide a harmless passage to the ground. The Empire State Building in New York gets struck 500 times a year without being damaged.

There are no real thunderbolts. Any explosion associated with a lightning strike happens because moisture in a tree, wall or soil vaporizes instantly. This causes explosive expansion which ejects the tree bark, mortar or earth violently. Objects said to be thunderbolts are meteorites or, more commonly, pieces of fulgurites – rodlike bits of dark glass produced when lightning passing through moist sand fuses the grains into "petrified lightning".

The temperature is five times that of the Sun and the air expands explosively with the deafening rumbling crash or crack of thunder.

Light travels a million times faster than sound so lightning is seen before thunder is heard. The distance in miles (kilometres) to lightning can be estimated by counting the seconds between lightning and thunder and dividing by five (three).

It took the brilliant insight of American Benjamin Franklin (1706-90) to prove that lightning was a form of electricity. In an experiment in 1752, he flew a kite made of a silk handkerchief into a thunderstorm and showed how an electrical spark flashed between a metal key tied to the end of the kite string and his knuckles.

He amused his friends by giving them electrical shocks but few realized how lucky they were until Professor Richmann of St. Petersburg died repeating the experiment in 1753. King Louis XVI of France once entertained his courtiers by using the experiment to pass an electrical shock through 200 monks who were all holding hands.

Lightning seeks out the point on the ground of least electrical resistance. Exposed hill tops, golf courses, sports fields and on open water are the worst places to be when lightning strikes. Many people increase their chances of being struck by sheltering beneath a tall tree, often holding a metal object such as an umbrella or golf club. When lightning strikes a tree it frequently jumps to a nearby person because the human body is a better electrical conductor.

People struck by lightning receive a severe electrical shock, which can stop the heart, and they may be badly burned, but prompt artificial resuscitation can revive the victim. The safest place to be during a thunderstorm is inside a building. If lightning does strike it follows the water pipes or electrical circuits and can side-flash to a person; for this reason telephoning should be avoided during thunderstorms. Enclosed vehicles are also relatively safe havens because the metallic body conducts the lightning safely around the occupants before earthing to the ground across the tyres.

Devils and twisters

Fast-rising thermals that generate and sustain a thunderstorm can sometimes be sent spinning to form the narrow column of violently rotating winds that is a tornado. A tornado begins as a funnel-shaped cloud protruding from the base of a severe thunderstorm which quickly extends to the ground. Its frenzied winds swirl at speeds often exceeding 100 mph (160 km/h) and occasionally reaching 300 mph (480 km/h).

The United States experiences 700–1,200 tornadoes, or twisters, each year, causing 50–100 deaths. In April 1974, a 19-hour "super outbreak" of 148 tornadoes swept across 13 states killing 315 people, injuring 6,142 and causing $600 million worth of damage.

The counterclockwise swirling winds of a severe tornado can destroy buildings, overturn vehicles, flatten mobile homes and uproot trees along a path of destruction often only a hundred yards wide. Buildings a few yards outside the tornado track remain unscathed.

The lifting power of some tornadoes is colossal, with vehicles, roofs, people and animals being lifted – and sometimes carried – considerable distances. In Minnesota, in May 1931, a tornado lifted five coaches from a moving express train killing one person and

injuring 57 others. In May 1986, in China, 13 schoolchildren were reportedly sucked up into a tornado and carried for 12 miles (19 km) before being dropped gently to the ground.

In the United States tornado warnings are issued once a tornado is sighted or weather radar has detected a tornado developing inside a thunderstorm. With a few minutes to impact, seeking shelter inside a sturdy building and keeping away from the windows where lethal missiles of debris may enter is the correct action to take. As a tornado struck Gillespie, Illinois, in March 1948 a woman took shelter in a small cupboard under the stairs. When the storm had passed she found the cupboard and stairway were all that remained of her house.

No tornado warning was issued to the small town of Saragosa, Texas, in May 1987. In a few seconds a tornado devastated its buildings, killing 40 and injuring more than 100 people out of a total population of only 185.

Tornadoes move too fast and too erratically for people to flee from them. In Wichita Falls, Texas, 26 out of the 43 people killed by the April 1979 tornado died in their cars as the tornado struck. One man tried desperately to hold on to his wife, but the suction of the tornado pulled her through the window.

Tornadoes come in many shapes and hues. They can appear as a narrow cylindrical column – as in this Nebraskan tornado – a broad cone shape, a thick amorphous mass, or a thin twisted rope. Broad tornadoes can consist of several smaller but intense mini-twisters that rotate about a common centre.

A tornado is made visible by water vapour condensing into cloud droplets in the low air pressure of the core. It is often a dirty white to grey or even blue grey but can be coloured exotically if the wind sucks up red or yellow sands. Some may be luminous at night if they have lightning in their cores.

Tornadoes strike in Argentina, Australia, Bangladesh, Europe, Japan, South Africa and the U.S. Thunderstorms spawn tornadoes most often in mid-latitudes as such storms form where winds converge from different directions thus encouraging upcurrents to spiral.

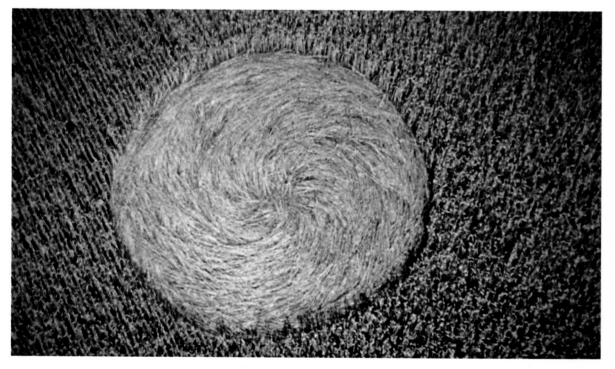

Many summer fields in southern England reveal near-perfect circles of flattened crops. Since 1980, when they were first investigated, nearly 3,000 crop circles have been found in fields in over 30 countries of the world.

Most are plain circles, between 5 and 30 yards (4.5 and 27 m) across, but some display rings and small surrounding circles. Crop circles are caused by localized whirlwinds which develop in the wind as it flows around hills and escarpments. Sometimes these whirlwinds wobble and begin to break up. As they do so they send a doughnut-shaped bulge in their spinning column plunging suddenly down, swirling the crop.

In the eye of the storm

To prevent overheating of the world's tropical oceans by intense summer sunshine, the atmospheric heat-exchange engine unleashes huge and powerful storms to carry surges of energy away from the tropics toward middle latitudes. As many as 100 violent tropical cyclones or hurricanes arise each year. As each storm moves away from low latitudes it releases energy comparable to a massive volcanic eruption. A day's energy output from one storm, if it could be harnessed, would power the entire industrial production of the United States for one year.

Tropical cyclones are complex affairs. Few tropical storms reach the hurricane stage with its distinctive calm clear eye, spiralling bands of thunderstorms and wind speeds in excess of 74 mph (118 km/h). Each year, meteorologists track 100 small storms developing off the west African coast between June and October but only 5 or 6 intensify to become life-threatening hurricanes.

Tracking hurricanes as they approach and threaten countries in and around the Gulf of Mexico is essential if their path of destruction is to be predicted accurately so that lives can be saved through evacuation. Although hurricanes move forward only slowly, around 15 mph (25 km/h), their track is often erratic. All too often, one community is evacuated following a hurricane warning, only for the storm to make landfall 50-100 miles (80-160 km) from where it had originally been predicted.

The National Hurricane Centre in Miami is improving the accuracy of its 24-hour landfall warning by about half a percent every year.

Vast amounts of moisture evaporate from the heated tropical oceans. As this hot moist air rises it condenses to form clouds. Air rushes inward at the surface to replace the rising air. This evaporates more moisture from the sea, transferring energy from the ocean to the atmosphere, which encourages cumulonimbus clouds to form and rainfall to intensify.

The winds strengthen and, in the northern hemisphere, swirl counterclockwise (clockwise in the southern hemisphere) around a central area, or eye, of the storm. Fast-rising air is thrown outward at the

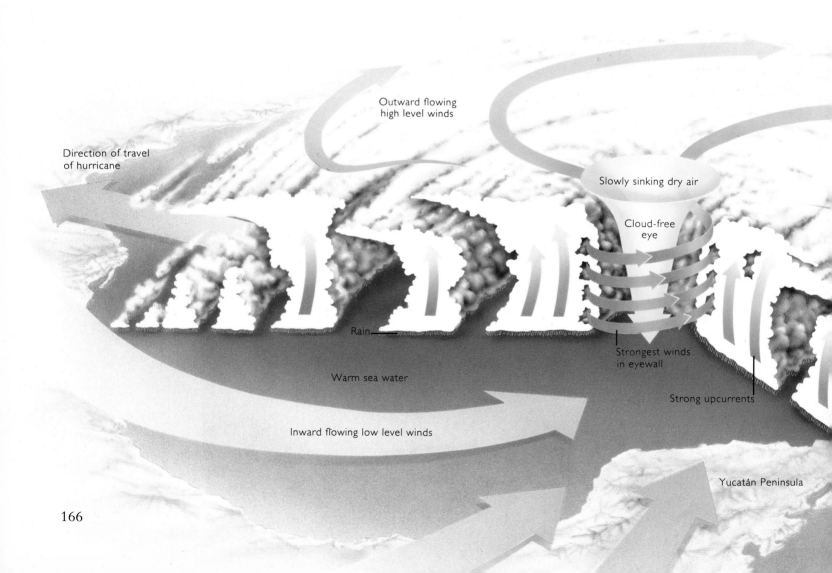

Outward flowing high level winds

Direction of travel of hurricane

Slowly sinking dry air

Cloud-free eye

Rain

Strongest winds in eyewall

Warm sea water

Strong upcurrents

Inward flowing low level winds

Yucatán Peninsula

166

top of the spiral bands of thunderstorms. Some turns into the eye and sinks slowly. It encounters lower air pressure, warms and dries. Any cloud in the eye evaporates and the eye becomes cloud-free.

Once the hurricane can sustain its input of energy from the ocean, the inflowing winds spiral faster. The eye becomes circular and smaller, typically 15–25 miles (25–40 km) across. Although the hurricane may be 300–900 miles (480–1,450 km) across, its destructive winds are usually restricted to within a radius of about 60 miles (100 km) of the eye.

Where tropical cyclones form

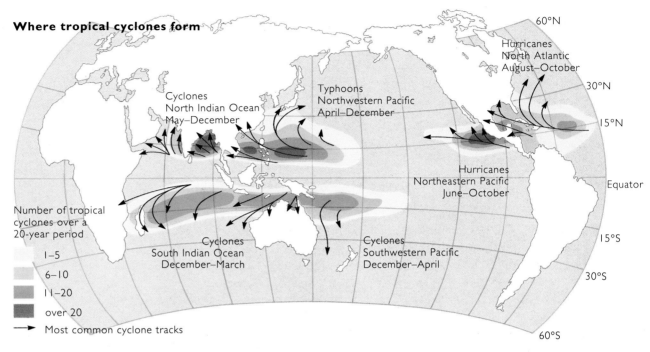

Hurricanes
North Atlantic
August–October

Cyclones
North Indian Ocean
May–December

Typhoons
Northwestern Pacific
April–December

Hurricanes
Northeastern Pacific
June–October

Cyclones
South Indian Ocean
December–March

Cyclones
Southwestern Pacific
December–April

Number of tropical cyclones over a 20-year period

1–5
6–10
11–20
over 20

→ Most common cyclone tracks

Florida

Cuba

Prevailing northeast trade winds

Tropical cyclones develop over the warm oceans during the hottest months of the year, when surface temperatures exceed 80°F (27°C). South America is free from these storms because its adjacent waters are relatively cool.

These cyclones do not form close to the equator because the planet's rotation force, needed to initiate the inward spiralling of the winds, is too weak. Once formed, tropical cyclones drift westward with the trade winds and eventually turn into the mid-latitudes as they decay.

Tropical cyclones are called hurricanes in the Atlantic after Hunraken, the Mayan god of winds. In the Indian Ocean they are simply called cyclones, meaning coiled serpents; and in the western Pacific they are known as typhoons after Tai Fung, Chinese for "great wind". The names may differ but their potential to bring death and destruction is the same.

In the eye of the storm/2

Aircraft fly through violent eyewall clouds to reach the eye of the hurricane to collect vital weather information, and satellites monitor the storm's erratic movement from space.

In September 1988, Hurricane Gilbert became the most powerful hurricane this century, creating a swath of devastation through the Caribbean islands, the Yucatán Peninsula and into Mexico. The lower the air pressure in the eye of a hurricane, the stronger the air pressure gradient between the outer and inner parts of the storm and consequently the higher the wind speeds. Hurricane Gilbert produced a record low central air pressure of 882 millibars, and its swirling winds reached sustained speeds of 175 mph (280 km/h) with gusts of 218 mph (350 km/h). Few structures are built to withstand the force exerted by such winds. People caught outside were tossed about like leaves and assaulted by a lethal barrage of flying missiles such as road signs, branches, roof panels and jagged shards of glass.

As Gilbert moved slowly forward, its ring of eyewall thunderstorms dumped vast amounts of rainfall on to the land, transforming streams into raging torrents and triggering mudslides. At Monterrey, Mexico, five buses carrying coastal evacuees were engulfed in turbulent flood waters, killing 140 people. In total, Hurricane Gilbert killed 320 people – 220 in Mexico alone and caused $8 billion worth of damage during its destructive lifetime.

Hurricane Gilbert began to die out as it crossed the Mexican and south Texan coastline and became cut off from its oceanic fuel supply. Although its winds diminished, its potential for destruction did not. Not only did the hurricane unleash torrential rains but the thunderstorms on the outer fringes of the storm spawned more than 20 tornadoes to create further destruction.

Worse tropical cyclones are yet to come. In the coming decades, these storms may become more frequent and more intense as global warming forces the atmospheric heat-exchange engine to operate at a greater capacity. Tropical oceans will exceed the critical surface temperature value of 80°F (27°C) over larger areas and for longer duration, which will result in an increased potential for storm generation.

Hurricane Gilbert struck Jamaica in September 1988 leaving 36 people dead and half a million – a third of the population – homeless, as buildings had their roofs torn off and walls toppled. Roofs, boats, vehicles and aircraft – as seen here at Kingston – were tossed about like toys. The country faced economic ruin as half the export crops of bananas, coffee and coconuts were destroyed and tourism suffered.

The low air pressure in the hurricane's eye raises the sea into a giant dome 1–2 feet (0.5 m) higher than the surrounding ocean. As this storm surge enters shallow coastal waters it rapidly increases in height. Gilbert's waves rose to 20 feet (6 m) as they swept across the low-lying shores of Jamaica and other islands, with widespread destruction and loss of life.

Hurricane Agnes struck the U.S. in June 1972 causing $4 billion worth of damage (left). Its torrential rains produced record floods, which required 250,000 people to be evacuated, and 117 lives were lost.

The greatest loss of life caused by a hurricane in the U.S. happened in September 1900 when the city of Galveston, Texas, was battered to the ground with 6,000 people killed and 5,000 injured.

Tropical cyclones striking the coastal areas of densely populated developing counties have an even greater impact. In April 1991 a cyclone with winds reaching 140 mph (225 km/h) generated waves 23 feet (7 m) high, which swept inland drowning 130,000 and injuring 460,000 people in Bangladesh.

HOW LONG CAN IT LAST?

How long can it last?

Not since the demise of the dinosaurs and many other creatures in a series of mass extinctions some 65 million years ago has the planet been exposed to such rapid environmental change on so wide a scale. If present trends continue, nearly one-third of existing plant and animal species will become extinct within the next 40 years, nine-tenths of them from the tropical rainforests. Restoring this lost biological diversity may take Earth millions of years to achieve – if indeed it ever can.

The population explosion of recent decades has resulted in a phenomenal rise in the consumption of Earth's natural resources including fossil fuels, minerals and fresh water. In turn this heightened consumption has caused gross pollution of the atmosphere, seas, rivers, groundwater and land.

Whereas it took a full century for world population to double from one billion to two billion between 1830 and 1930, it took just under half a century for it to double again, reaching four billion in 1975. Since then billions seem to have been increasing by the decade: five billion in 1987, six billion projected for 1997 and even 10 billion by 2040.

Most population growth is taking place in the developing countries, but it is the extravagant consumption of resources by developed nations that causes the greatest share of environmental degradation and planetary stress. Although developed countries comprise only 25 percent of the world's population, they consume 80 percent of the Earth's resources.

In an attempt to keep pace with the ever increasing number of mouths to feed vast areas of natural vegetation have been cleared

and agriculture has expanded into many vulnerable areas. Overuse of land has triggered desertification, in which a loss of plant cover and soil fertility ultimately leads to loss of the soil itself as it is blown away by winds and washed away by rain and rivers. Each year 23,000 square miles (60,000 sq km) of land – almost the size of Sri Lanka – turn to desert and nearly one-third of Earth's land area is now threatened with desertification.

Poorly managed irrigation schemes are ruining productive land. Half the world's irrigated lands are either waterlogged through lack of adequate drainage or have saline soils because the water evaporates, leaving a salt deposit. Rivers feeding into irrigation schemes or dammed to generate hydroelectricity are suffering too.

The Nile Delta in Egypt is no longer replenished by river sediments because of the Aswan High Dam and is being eroded by the sea. Attempts to grow cotton irrigated by Aral Sea water have shrunk the lake to one-third of its original size, its waters are saline, its fish dead.

Intensive modern farming relies on large and regular applications of pesticides, herbicides and fertilizers. Toxic chemicals and nitrates are flushed into rivers, polluting drinking water supplies and killing river life. Greater water contamination is caused by industrial waste and sewage generated in and around urban areas. Here not only is the

An all too familiar scene on the urban landscape (previous page). A factory belches out waste products into the Earth's atmosphere, making the air thick with pollutants which will destroy nature's delicate balance.

water contaminated but the air is too. A total of one-third of the world's people breathe air that is unhealthy.

One of the greatest tragedies of recent decades has been the destruction of large areas of the tropical rainforests of Brazil, Colombia, Indonesia, Thailand, Laos, Central America and parts of Africa. These highly productive ecosystems are being felled for their timber or burned and cleared to make way for cattle ranching, mining of iron and aluminium, farming and the construction of dams to generate hydroelectricity.

Roads are bulldozed into the heart of the Amazon to encourage colonists – landless families from outside the region – to farm the cleared land, even though only 3 percent of Amazonian soil is considered sufficiently fertile to sustain farming after one harvest. One-seventh of the Amazon forest has been lost already while 45,500 square miles (120,000 sq km) of forest – almost the size of England – are being destroyed each year.

The consumption of fossil fuels like coal, oil, lignite and natural gas for power generation, industrial processing, domestic heating and cooking, and for powering the 600 million motor vehicles on the roads of the world, generates considerable air pollution. Widespread contamination of rain and snow by sulphur dioxide and nitrogen oxides is making forest soils and the waters of upland lakes acid, resulting in the deaths of trees and fish in many regions.

Air pollution is posing problems on a global scale as chlorine-containing

compounds upset the natural ozone balance in the stratosphere and trigger ozone destruction. Each spring the ozone layer above Antarctica is depleted by over 50 percent, which allows lethal ultraviolet radiation to reach the Earth's surface, damaging phytoplankton with a profound knock-on effect higher up the food chain. The Arctic, Europe and North America are now under threat from massive ozone depletion too.

Fossil fuel burning, cattle ranching and rice cultivation are adding heat-trapping greenhouse gases to the atmosphere which threaten to create a "hothouse" Earth in the coming decades. Rising global temperature will swell the seas, causing world sea levels to rise, and alter regional rainfall

and storm patterns. Considerable strain will be placed on agriculture, water supplies, coastlands and natural ecosystems if they are to cope with such rapid changes.

What has taken hundreds of millions of years to evolve is being destroyed in a matter of centuries or decades. If the trend continues, the future of our living planet lies in the balance. Today a billion people live in abject poverty, 600 million are on the edge of starvation and 50,000 children under five years old die from hunger and disease every day.

Unless prompt action is taken to halt the environmental degradation we are causing, there is no doubt these statistics of human misery will continue to increase. The warning signs are clear. Whether they are heeded and acted upon by both individuals and nations remains to be seen.

City climes

Replacing farmland, forests and grassland with car parks, factories, houses and roads alters the temperature, humidity, airflow and air quality of an area. As a city expands and building densities increase these differences are heightened and the city creates its own distinctive local climate, or microclimate.

City centres at night can be as much as 9–18°F (5–10°C) warmer than the surrounding countryside. A heat island develops not only because of heat released by industry, vehicles, central heating and air conditioning systems but also because asphalt, brick and roofing materials absorb large amounts of sunlight during the day. At night the urban fabric releases this heat which partly balances night-time cooling.

During cold winters this warm climate dome is welcomed by residents, but on hot summer nights sleeping is almost impossible. Temporary alleviation can occur in the outer suburbs because the heat island lowers surface air pressure slightly in the city centre, which draws in a cool breeze from the countryside.

Urban warmth and the irregularity of the city skyline promote cloud growth and rainfall, even triggering thunderstorms. Rainfall increases of 10–30 percent can extend up to 40 miles (65 km) downwind of some cities such as London, New York and Paris. Although the extra rain is welcomed by farmers, it is often highly acidic and contaminated by city pollution, and since much of it arrives in heavy storms, it causes soil erosion and floods.

When rain falls in urban areas it is quickly channelled into drains, and then into a river, often within hours. In rural areas, however, rain soaks into the soil, produces little surface runoff, and reaches the river in a matter of days or weeks. Urban concentration of runoff means river flow after rainstorms increases abruptly, causing floods.

Cities may provide inhabitants with the living and working conditions they need but they also have adverse effects on the environment. These problems can be tackled by attention to building design and choice of materials, street layout, type of heating systems, transport networks and by curbing pollution.

Noxious smog envelops Manhattan, New York City, during the Friday evening rush hour as tens of thousands of slow-moving vehicles' exhausts belch out health-damaging pollutants. Headache-causing carbon monoxide, asthma-triggering nitrogen oxides, cancer-forming benzene, brain-damaging lead, eye-stinging ozone and dirty soiling particulates are some of the pollutants found in the chemical cloud cocktail called smog.

Smogs develop in peak traffic hours in cold or hot weather when heating or air conditioning systems are turned up. They are also evident on calm days when winds fail to mix or disperse the pollutants. In cities such as Athens, Berlin, Los Angeles, Mexico City, New York, São Paulo and several in eastern Europe, smogs are so bad that a smog alert is issued warning: "breathing endangers your health".

Asthmatics, pregnant women, the elderly and those suffering from heart and lung diseases are advised to stay indoors. Healthy people are told not to exercise because deep breathing during strenuous exercise increases the pollutant dose they receive.

Three smog alert levels have been adopted by city authorities. The first indicates health is at risk, while the second and third levels aim to achieve voluntary and then compulsory reductions in emissions by encouraging car sharing or banning cars from city centres and by curbing industrial activity. In Santiago, Chile, the smog was so bad in 1988 that authorities sprayed it with water and detergent from aircraft to try to wash out the pollutants.

Acid rain

Dying fish in European and North American lakes first alerted people to the alarming effects of acid rain in the 1960s and 1970s. By the 1980s trees in some coniferous forests began to display signs of acid damage, with needles shedding or turning yellow and crowns thinning. Tree surveys have since revealed a severe problem in Britain, Czechoslovakia, Germany and Poland.

Rain is naturally slightly acidic, since it contains carbon dioxide, but increased air pollution from industry and vehicles has contaminated rainfall so much that some is as acid as lemon juice and 500 times more acidic than unpolluted rain. This is extreme, but even averaged over the year, parts of Europe and North America experience rain 10 times more acidic than usual.

Acid rain leaches vital nutrients such as magnesium, calcium and potassium from the soil depriving the trees; it even leaches nutrients from the leaves or needles. The loss of calcium can have effects right through the food chain: insect–eating birds may lay eggs with thin shells which are easily damaged.

In Sweden acid rain leaches copper from water pipes, contaminating water supplies and causing illness in children. It even tints laundry, baths and blond hair green. And acid rain is also eating away the stonework of cathedrals and irreplaceable historical monuments.

The political wrangles between nations in the 1980s over the need for action on acid rain have largely been settled. All parties now accept the need to reduce pollution. The next step is to increase the small pollution reductions already agreed to the larger reductions, as much as 90 percent, which environmentalists argue are essential if acid rain damage is to be halted.

Regions of the world suffering most from the effects of severe acid rain are those areas with sensitive soils – those lacking adequate amounts of acid-neutralizing compounds – located in or near areas of large pollution emissions.

Many developing nations, especially those in the tropics, are experiencing rapid urban-industrial growth and increased air pollution. It is likely that countries such as Brazil, China, India, Malaysia, Nigeria and Venezuela will face a severe worsening of acid rain in the near future. Rain of pH 4 has already been experienced in south China.

Burning fossil fuels in factories, power stations and vehicles releases sulphur dioxide and nitrogen oxides. These gases have an adverse effect on crops, trees and buildings locally, but large amounts are also blown by the winds for many hundreds of miles. On route they are transformed by the Sun's rays into sulphates and nitrates.

These pollutants resist being deposited on to the ground when dry and are removed from the atmosphere only by rain or snow. They are absorbed into clouds and form sulphuric and nitric acid which later produce acidic precipitation.

Mountain regions receiving acid rain or snow often consist of granite and other igneous rocks which produce thin soils lacking acid-neutralizing chemicals such as calcium (found in limestone). Acid rain thus poses severe problems for the forests, crops, lakes and people in these regions.

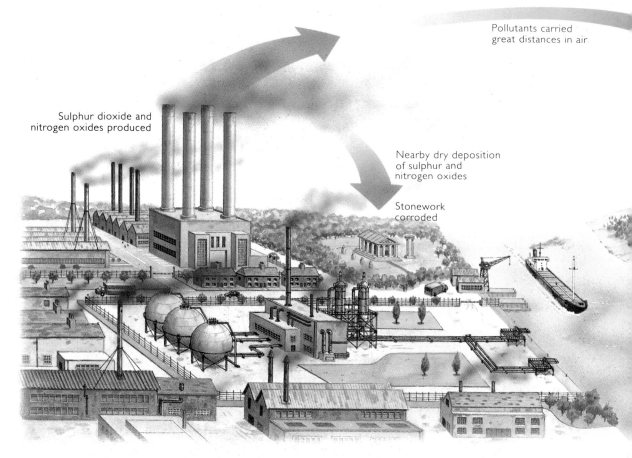

Pollutants carried great distances in air

Sulphur dioxide and nitrogen oxides produced

Nearby dry deposition of sulphur and nitrogen oxides

Stonework corroded

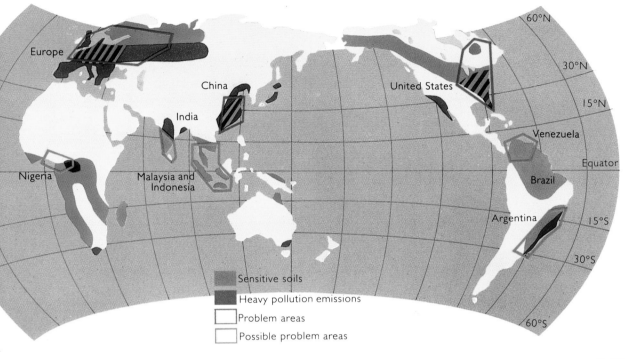

Sensitive soils

Heavy pollution emissions

Problem areas

Possible problem areas

The pH scale measures the acidity and alkalinity of liquids, with pH 7 being neutral (distilled water). Each unit on the scale increases acidity tenfold: rain with pH 3 is 10 times as acidic as pH 4 and 100 times more than pH 5.

Lakes have a pH close to neutral because minerals such as calcium, washed into lakes from the soil, neutralize natural rain. However, the buffer may be unable to absorb any increased acidity in rain, often with devastating effects on the lake fish.

Aluminium may also be released from soil by acid rain and washed into lakes where it causes fish to overproduce a sticky mucus which clogs their gills.

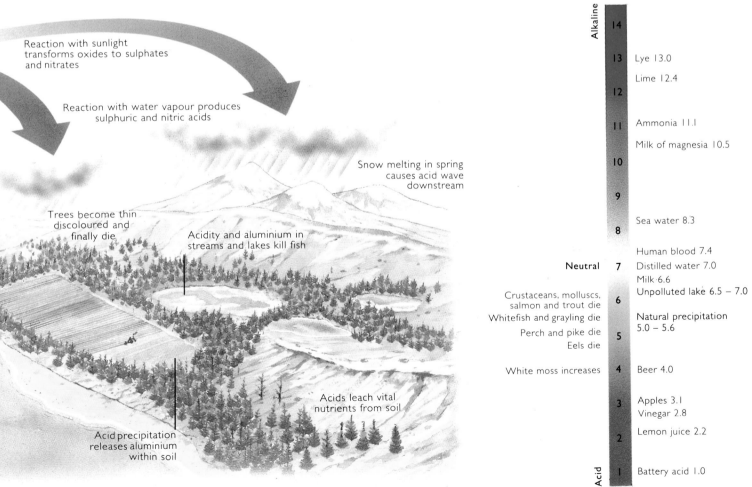

Reaction with sunlight transforms oxides to sulphates and nitrates

Reaction with water vapour produces sulphuric and nitric acids

Snow melting in spring causes acid wave downstream

Trees become thin discoloured and finally die

Acidity and aluminium in streams and lakes kill fish

Acids leach vital nutrients from soil

Acid precipitation releases aluminium within soil

Alkaline	14	
	13	Lye 13.0
		Lime 12.4
	12	
	11	Ammonia 11.1
		Milk of magnesia 10.5
	10	
	9	
	8	Sea water 8.3
		Human blood 7.4
Neutral	7	Distilled water 7.0
		Milk 6.6
		Unpolluted lake 6.5 – 7.0
Crustaceans, molluscs, salmon and trout die	6	
Whitefish and grayling die		Natural precipitation 5.0 – 5.6
Perch and pike die	5	
Eels die		
White moss increases	4	Beer 4.0
	3	Apples 3.1
		Vinegar 2.8
		Lemon juice 2.2
	2	
Acid	1	Battery acid 1.0

Changing landscapes

Natural forces are constantly at work shaping and altering the landscapes of the Earth. Long-term climate changes, thought to be the result of subtle shifts in the Earth's orbit and slight changes in the amount of energy radiated by the Sun, expose the land to fluctuating temperatures and rainfall. And the slow geological processes of plate tectonics, which move landmasses from icy to temperate to tropical regions and back again, also expose the land to different climates over millions of years.

As part of that geological process the Earth's plates collide and pull apart creating hills, mountains and new land out of earth and fire which are then exposed to the other two elements – air and water. These, in the guise of wind, rain, snow and ice, are the elements that work to level the land again through the processes of weathering, erosion and deposition of sediments. They create the landscapes with which we are familiar.

But there are forces at work preventing weathering and erosion: once established on a new land surface, vegetation resists erosion by meshing the soil in a living net of roots. This, and the physical barrier of vegetation above ground, holds the soil in place in the face of rain which would otherwise wash loose material downhill. Vegetation also creates microclimates which soften the extreme effects of climate.

In fact, in the context of the inexorable changes brought about by large-scale geological and climatic variations, the effect of vegetation is a stabilizing one – it maintains the shape of the land. But plants are not the only living things to colonize the landscape. Creatures of all kinds live in and off the various vegetation types. But just one species appears to be having a profound effect on world landscapes: in the last few thousand years, and at a seemingly accelerating pace, people have been making large-scale changes.

These effects have largely been caused by changes to vegetation. Destruction of rainforest in the tropics by cutting down trees to sell the wood or to clear space for crops or cattle is the most obvious and well-publicized example.

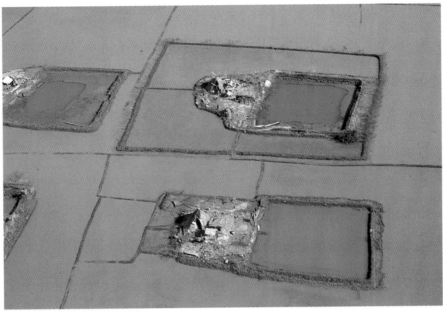

Felling Amazonian rainforest (far left) raises questions about human interactions with the environment. On the one hand it would seem that destroying an ecosystem that has evolved over millions of years to make use of poor soils and high rainfall is risky. It can lead to loss of soil fertility and rapid erosion. But on the other hand, to a poor farmer or timber worker, it provides space to grow crops for a family that might otherwise starve or, when logs are sold abroad, it provides an income with which to buy food.

In Bangladesh, the inhabitants have to cope with the torrents of water that drain off the Himalayas following the monsoon rains. By careful water management (above), they can prevent soil being washed away and encourage the deposition of new, fertile banks of silt. The silt comes from deforested regions upstream where soil is eroded and water is released quickly in a flood.

A hillside before and after deforestation (left). Some rain falling on a forest is evaporated or transpired back into the atmosphere. Rain that reaches the ground is absorbed by the roots or held in the soil. This soil water is then released slowly into rivers.
After deforestation gullies form down which soil and water rush. In heavy tropical rain, the land erodes and floods. Landslides occur and the cleared land – often used for cattle grazing – rapidly degrades. Where crops are grown, the rate of erosion is less but the land is not able to support intensive agriculture for long.

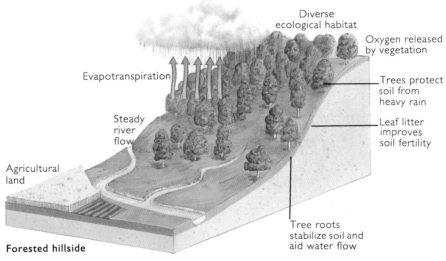

Diverse ecological habitat
Oxygen released by vegetation
Evapotranspiration
Trees protect soil from heavy rain
Steady river flow
Leaf litter improves soil fertility
Agricultural land
Tree roots stabilize soil and aid water flow
Forested hillside

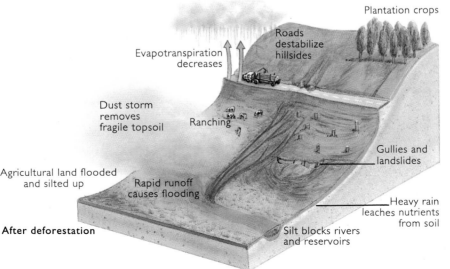

Plantation crops
Roads destabilize hillsides
Evapotranspiration decreases
Dust storm removes fragile topsoil
Ranching
Gullies and landslides
Agricultural land flooded and silted up
Rapid runoff causes flooding
Heavy rain leaches nutrients from soil
After deforestation
Silt blocks rivers and reservoirs

179

Changing landscapes /2

Once the vegetation has been cleared, water washes away the soil and erodes the landscape. Later the soil is deposited in lakes or in deltas where it may interfere with the natural formation of water channels.

Changes to vegetation also have other, more subtle, effects. One of these is exemplified by the creation of the "dust bowl" in the American Midwest in the 1930s. First the grassland was removed, and this was followed by the repeated growing of a single crop year after year. This agricultural practice alone was to blame for the terrible conditions experienced by farmers. The soil was laid bare by drought, which then prevented crops being grown, and the wind removed the topsoil.

Intensive agriculture in many parts of the world is today responsible for the loss of a great deal of fertile topsoil. But it is not only alterations to the vegetation cover that change the landscape. The building of dams for hydroelectricity and the overextraction of water from rivers for irrigation alter the way that rivers carry sediments. Large-scale mineral

extraction causes subsidence and spoil heaps. But there are some man-made changes to the landscape that seem to be of benefit, however, and these include the irrigation of arid regions to grow crops, and the planting of hardy trees to provide windbreaks. But there are always pluses and minuses to every modification.

The engine that drives the continued assault on the environment is population pressure. As the number of people rises, there are more mouths to feed and, where sheer survival is not the pressing problem, expectations for improvements in living standards have to be realized. It is hard for a herder in the marginal desert lands not to overgraze the land, thereby hastening desertification, when members of the family are hungry.

In many cases it is difficult to tell precisely what has been the key factor behind a change. Humans are, after all, just one component in a complex system. However, what is certain is that the consequences of landscape changes are at best unpredictable. At worst they could be disastrous.

By agriculture, people have shaped the landscape to suit their needs. These rice terraces in the Philippines, built at least 2,000 years ago, are carefully maintained to make the most use of the available fertile land in this hilly region.

Streams are diverted and water collects in the stepped fields. The rice terraces are estimated to be about 14,000 miles (22,500 km) long, and this increases annually as farmers continue to encroach on the mountainside, using the rock to build retaining walls around small areas of rice beds. This is a labour-intensive way of catering to the food requirements in a densely populated region.

Irrigating the desert in Egypt creates a massive fertile circle of various shades of green. Without the precious resource of water the crops could never have been grown.

The supplies have to be secure and constant and this means that water is either diverted from a dammed river or extracted from deep below the ground.

Deserts on the march

There is no more precarious existence than scratching out a living in the arid lands that border a desert. But many millions of people live in these regions. In fact, up to 135 million people in about 100 countries in Africa, the Indian subcontinent and South America face the problem of desert encroachment.

About one-third of the planet is classified as arid or semi-arid. The key to this is rainfall: deserts have less than 10 in (250 mm) of rain a year; semi-arid regions about twice that amount. According to some authorities, an average of 40 square miles (100 sq km) of the Earth's surface become desert every day. The term used to describe this ever advancing threat is desertification.

The spread of desertlike conditions is due to human influence or to a change in climate. The current speed of the change is thought to be caused by a combination of human activities, both on a local and global scale, and a run of extremely bad drought years. On a global level the effects of the increase in greenhouse gases may be contributing to the rising number of years when drought prevails.

At the local level removing vegetation for firewood, intensive grazing of livestock and intensive crop cultivation on poor soils leads to a breakdown in the soil-binding root systems of vegetation cover. Thus the soil has nothing to hold it in place and it blows away or is washed away by the infrequent rain. With no soil, there can be no crops and the surface cannot retain any rain. The desert advances in patches into once fertile territory and people suffer increasing hardship as famine sets in.

In the Sahel in Niger, Africa, millet is being grown in an attempt to stem the Sahara Desert and to feed the starving population of this once relatively good agricultural area. This is one of the areas most severely affected by drought, and the southward advance of the desert has increased sixfold in recent years.

Various methods are being tried in an attempt to halt the march of desertification. In Algeria trees are being planted bordering the Sahara and it has been suggested that this green ring could be extended around the entire desert fringe.

The Sahara was once a fertile land of trees and grassland, with fields of grain cultivated and harvested by the Romans. This rich land once echoed to the calls of abundant wildlife. Now it responds to the call of the vulture.

Almost 20 percent of the Earth's land surface is already desert and another 13 percent is semi-arid. The map shows these areas together with those that are most at risk from desertification.

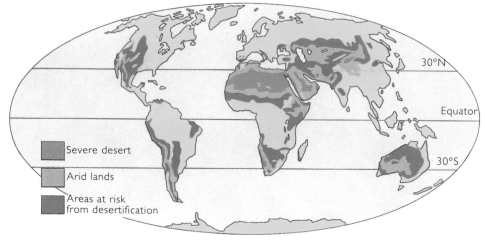

Severe desert

Arid lands

Areas at risk from desertification

30°N

Equator

30°S

Controlling the weather machine

From understanding the weather machine to being capable of controlling it is a giant leap. In earlier times fine-tuning the weather was restricted to appeals to the weather gods. Religious sacrifices, ringing of church bells, prayers and rain dances were tried but without great success. In the 1940s, however, scientists discovered that seeding clouds with chemicals such as silver iodide and dry ice could trigger rain or snow from some clouds.

Cloud seeding chemicals behave like ice crystals. They cause cloud water droplets to evaporate and transfer to the seeding nuclei. These grow in size to become large ice crystals which fall from the cloud as snow, sometimes melting to arrive as rain. Rainfall increases of 25 percent have been claimed for seeding attempts but these have not been without problems. When a disastrous flood befell Rapid City, South Dakota, in 1972 cloud seeding nearby was blamed, possibly unjustly.

Weather control experiments are wide-ranging. Russians claim Moscow is kept free of snow by seeding snow clouds with silver

Windstorms such as this cyclone in the Bering Sea off the Kamchatka Peninsula (shown in computer-enhanced false colours) cause more deaths and destruction than any natural hazard.

Seeding hurricanes with chemicals to produce rain has been tried in an attempt to weaken their winds. In 1947 a storm moving from the east coast of the U.S. was seeded but it turned around, headed for Georgia

and caused severe damage. Apart from success in 1969 in weakening Hurricane Debbie's winds by 30 percent for a few hours it was decided that hurricanes were not as easy to control as was once thought.

iodide. They even spray cement dust on top of clouds, suppressing their upcurrents, as a cheaper method of reducing snow. American scientists use small rockets with trailing wires to earth thunderstorms if there is any risk of lightning striking the space shuttle during launch. In wine-growing regions in Europe, hailstorms are overseeded to encourage formation of large numbers of small hailstones which are far less damaging than large hail.

Fog causes major disruptions to airports. Seeding fog can remove it as can heat from an array of engine exhausts as tried at Orly Airport, France. In the 1960s large nylon-filament brooms were placed along a New Jersey highway to remove fog. The brooms revolved, collecting fog droplets on their bristles. The experiment was abandoned when too many startled drivers had accidents.

Success in controlling the weather is limited and variable from place to place. Since the consequences of our interference are unknown, our efforts should focus on improving weather forecasting, not trying to be weather gods.

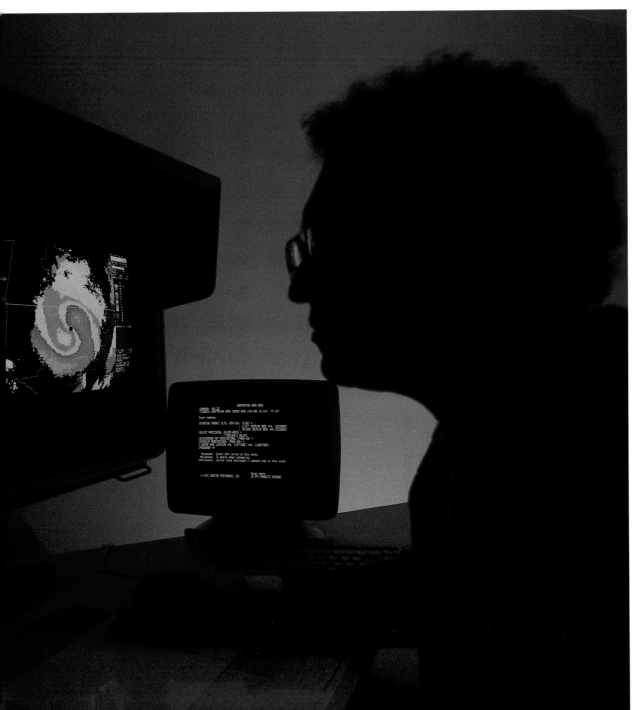

Advanced technology plays a vital part in modern weather control and forecasting. High-speed computers are essential to make sense of the vast amounts of weather information received from radar, satellites, weather balloons, automatic ground stations, ocean buoys and manned weather stations.

Such information arrives as instantaneous, or real-time, data. Satellites looking down from above can track the path taken by an erratic storm, while radar can see inside a storm and monitor it.

Radar is used to identify storms that are beginning to form large hailstones. Once a potentially dangerous hailstorm has been identified, rockets are fired to explode inside the storm releasing silver iodide crystals.

These act as nuclei for hailstone formation and encourage many small hailstones to form rather than fewer large and more damaging ones. The small hailstones either cause little damage to crops or melt as they fall to the ground and arrive as welcome rainfall. French and Russian scientists claim great success at this form of control.

The ozone layer

Human activities are severely damaging the ozone layer, which protects life on Earth from the harmful effects of the Sun's ultraviolet rays. Chlorofluorocarbons (CFCs) and other chemicals are destroying this high-level sunscreen and allowing increasing amounts of biologically damaging radiation to flood through to the Earth's surface.

Ozone depletion is worsening over the Arctic and the heavily populated northern middle latitudes, but Antarctica is the region that is worst affected. Each austral spring, virtually all the ozone disappears from between 8 and 14 miles (13 and 22 km) in the Antarctic stratosphere. This Antarctic ozone hole, $1\frac{1}{2}$ times the area of the United States in size,

affects ozone below 14 miles (22 km). Above this it is largely unaffected so in fact the "hole" is a thinning not a gap.

Ozone, a gas formed from three oxygen atoms rather than the two atoms forming "normal" oxygen gas, is found throughout the atmosphere. Since oxygen first appeared in our atmosphere there has been an ozone layer which has been in a state of dynamic balance for millions of years. Ozone is continually created and destroyed but the total has remained about the same except for small fluctuations caused by the Sun's activity or volcanic eruptions.

The situation changed in the 1960s when chlorine-containing chemical compounds such as CFCs began to be emitted in large

Chlorine levels in the atmosphere need to be reduced to below 2 parts per billion (ppb), the level at which the Antarctic ozone hole developed in the late 1970s.

The graph predicts the effects on chlorine levels of the 1987 world agreement to halve CFC emissions by 1999 (A); the revised 1990 agreement to ban all CFCs by 2000 (B); and the proposal to prohibit all chlorine emissions by 2030 (C). Clearly, the Antarctic hole will persist well into the next century.

Ultraviolet radiation
triggers complex competing chemical reactions which create and destroy ozone in the stratosphere. A natural balance has been kept for millions of years but a rise in pollution has tilted the balance toward destruction.

Chlorine from CFCs is the worst culprit, and it would deplete ozone more rapidly were it not for methane and nitrogen compounds tying it up into harmless compounds, namely hydrochloric acid and chlorine nitrate. Only where ice clouds occur in the ozone layer, as over Antarctica, does the chlorine escape being trapped into inactivity or when volcanoes produce tiny sulphuric acid droplets in the stratosphere, altering the chemical reactions.

Bromine depletes ozone but occurs in small amounts so far, while nitric oxide can also scavenge ozone. Carbon dioxide, methane and nitrous oxide are all greenhouse gases. They trap heat and warm the lower atmosphere, but this cools the stratosphere speeding up the chemical reactions which deplete the ozone.

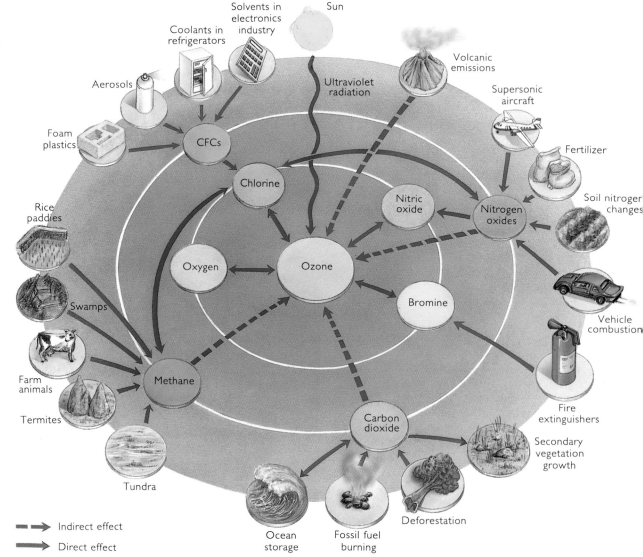

Ultraviolet radiation · Solvents in electronics industry · Sun · Coolants in refrigerators · Volcanic emissions · Aerosols · Supersonic aircraft · Foam plastics · CFCs · Fertilizer · Chlorine · Nitric oxide · Nitrogen oxides · Soil nitrogen changes · Rice paddies · Oxygen · Ozone · Swamps · Bromine · Vehicle combustion · Farm animals · Methane · Termites · Fire extinguishers · Carbon dioxide · Secondary vegetation growth · Tundra · Ocean storage · Fossil fuel burning · Deforestation

- - → Indirect effect
— → Direct effect

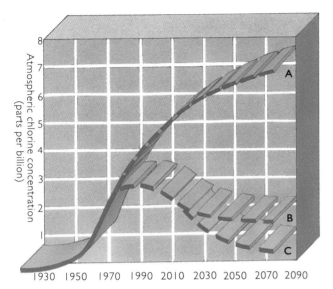

quantities. CFCs were used as propellants in aerosol sprays, as cooling fluids inside the pipes of refrigerators and air conditioning systems, as foam-blowing agents, and as solvents for cleaning electronic circuit boards.

CFCs remain in the atmosphere for 60 to 120 years and enter the stratosphere where sunlight releases the chlorine. Chlorine acts as a catalyst, upsetting the natural chemical balance in favour of ozone depletion. Each chlorine molecule converts ozone into oxygen without being destroyed itself and can scavenge as many as 100,000 ozone molecules.

The Antarctic stratosphere is particularly vulnerable to ozone depletion by chlorine because of its special weather conditions.

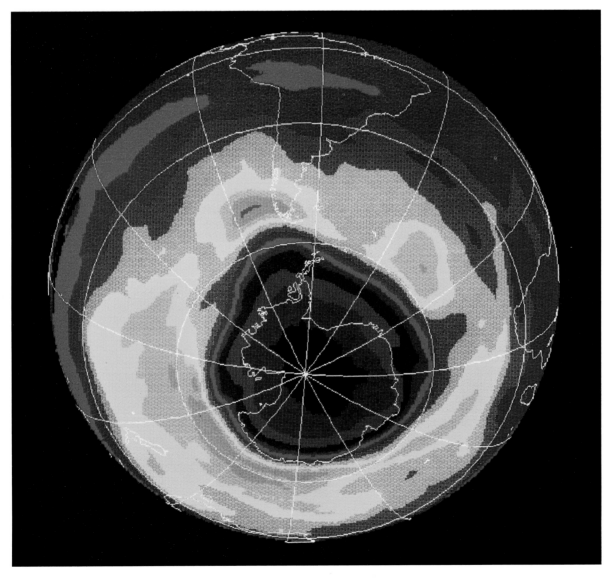

The ozone hole over Antarctica on October 3, 1990, ten years after it was first observed. This satellite image map was made by the Total Ozone Mapping Spectrometer (TOMS) on board the American Nimbus-7 weather satellite, and represents the maximum depletion during the Antarctic spring of 1990. The "hole" appears here as the pink, purple and black areas covering Antarctica, outlined in white, and beyond.

The ozone layer/2

During the cold dark austral winter Antarctica is isolated from the rest of the atmosphere by a powerful swirling vortex of westerly winds. Temperatures fall to $-121°F$ ($-85°C$), cold enough for ice particle clouds to form. The cloud particles trap nitrogen compounds stopping them from neutralizing chlorine (neutralizing tends to occur in warmer cloud-free parts of the global stratosphere). The ice particles also convert chlorine into a highly reactive substance enabling it to attack ozone molecules unopposed when the Sun returns in September.

The Arctic stratosphere is less vulnerable to chlorine attack. Because it is warmer, fewer clouds form, and it is not isolated by a swirling wind vortex. However, during winter there are small regions cold enough to form ice clouds and these do activate the chlorine. Winds sweep these chemical pockets into the middle latitudes, where they meet sunlight, so triggering rapid ozone destruction.

The consequences of ozone loss are profound. A 10 percent decrease in ozone leads to a 20 percent increase in ultraviolet radiation reaching the Earth's surface which may increase the number of cases of skin cancer in Caucasians by 40 percent. Increased radiation of this type causes eye cataracts, the single greatest cause of blindness in the world. It also suppresses the immune system, making the body less resistant to infectious diseases and reducing the effectiveness of vaccinations.

Since ultraviolet radiation penetrates the upper layer of the oceans, higher levels damage zooplankton and phytoplankton. In turn, this will affect krill, the tiny animals that feed on phytoplankton, and then fish, seabirds, penguins, seals and whales.

World nations have agreed to ban CFCs and some other ozone-depleting chemicals by the end of the century. However, since these chemicals persist for decades, ozone depletion will continue. It seems that acceptance of unrestrained chemical emissions by one generation has meant that the unwelcome consequences have to be borne by the next. Whether the Earth's support systems can nullify the worst effects of this human mismanagement of the planet remains to be seen.

188

As the ozone layer continues to thin, and ultraviolet radiation streams through to bathe the Earth's surface, the risks of severe sunburn, skin cancers, infectious diseases and damage to the eyes are rising dramatically. In the future the pleasures of sunbathing without adequate protection against the increasing levels of ultraviolet radiation need to be weighed very carefully against the growing risks of ill-health.

Hothouse Earth

Children born in the next century will probably grow up in a world much warmer than it is now. They could face a future of shifting rainfall patterns, violent storms and rising seas. Signs of a changing climate are evident already with temperature having risen by 0.9°F (0.5°C) this century. By 2050 this warming could be by as much as 4°F (2°C), creating a world hotter than it has been for 120,000 years, when modern humans, *Homo sapiens*, evolved.

Global warming is happening because human activities are pumping vast amounts of greenhouse gases including carbon dioxide, chlorofluorocarbons (CFCs), methane, nitrous oxide, low-level ozone and even water vapour into the atmosphere. As concentrations of these gases escalate, they trap more and more of the heat given off from the Earth's surface which normally escapes into outer space.

The degree of warming will vary: continental interiors and high latitudes will warm most. Climate will also change, with some areas becoming drier, others wetter. The United States corn belt could suffer drought recalling the hardship of farmers in the 1930s "dust bowl" era. A fall in grain yields would profoundly affect the many countries which rely on American exports to satisfy food demands.

Not all places will suffer. The growing season will lengthen in high latitudes, benefiting Canada, Iceland, Norway, Scotland and Japan. Since plants need carbon dioxide for photosynthesis, higher concentrations will help them grow faster and yield more. But enhanced growth may not mean increased protein content so insects may consume more of the larger, but less nutritious, plants.

With increased carbon dioxide levels, weeds may flourish at the expense of crops, leading to a decline in crop yields. Pests and disease thrive in warmer conditions. To comabt these, crops may have to be shifted into areas with a more suitable climate, new crops introduced and water conservation implemented. Increased irrigation will be needed but flow in the Colorado River, source of many irrigation systems in the United States, is expected to fall by a quarter in the next 50 years.

The Earth's surface and clouds absorb sunlight and are warmed. They then emit heat which passes out into space. Some gases, acting like a greenhouse, trap part of the outgoing heat and send it back to the surface. This keeps the surface warmer than it would otherwise be, like an insulating blanket.

The natural greenhouse effect is vital for human survival but our activities are increasing levels of greenhouse gases to an unacceptable degree. This is forcing world temperature to levels not experienced for thousands of years. The graph below shows the projected degree of warming if gas emissions continue to increase (A), slow down (B) or are reduced greatly (C).

Current enhancement of the greenhouse effect is due to carbon dioxide (half the warming) and CFCs (a quarter), plus methane, ozone, nitrous oxide and water vapour.

Carbon dioxide is emitted when fossil fuels or grassland are burned and when rainforest is destroyed. Methane is released by bacteria in cud-chewing animals, from gas pipe and coal field leaks, from decaying organic matter in swamps and paddy fields, and when vegetation is burned.

Water vapour is added when fuels are burned and by irrigation, which increases evaporation. Low-level ozone is created in urban areas by photochemical reactions on hydrocarbons and nitrogen oxides released by vehicles and industry. Ozone in the stratosphere filters out lethal ultraviolet radiation but low-level ozone is unwelcome because it traps heat. Nitrous oxide is produced from animal wastes, nitrate fertilizers, and burning of vegetation and fossil fuels.

CFCs are the most worrying gases because each molecule traps heat 13,000 times more effectively than carbon dioxide. CFCs are used as propellants in aerosols, coolants in fridges and air conditioners, insulating gases in foam plastics, solvents, and sterilants. The CFC world ban in the 1990s was vital to attempts to slow the greenhouse effect but CFCs persist for many decades.

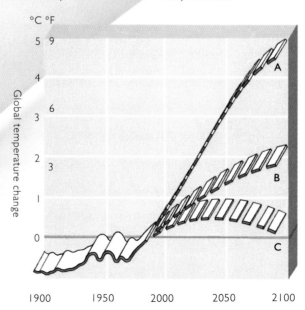

°C °F

Global temperature change

1900   1950   2000   2050   2100

Radiation from Sun

Energy reflected back to space

Radiation to space

Radiation absorbed by atmosphere and reradiated to Earth

Radiation absorbed by Earth's surface

Radiation to atmosphere

Reradiated to Earth

CFCs

Carbon dioxide

Nitrogen oxides

Methane

Ozone

Hothouse Earth / 2

Global warming is causing the surface layers of the oceans to warm and expand. World sea level has risen already by 4 inches (10 cm) this century and is expected to rise by 10–36 inches (25–90 cm) in the next because of ocean swelling and water added from melting ice sheets and glaciers. If the West Antarctic or Greenland ice sheet melted – which would take a few centuries – a world sea level rise of 16–23 feet (5–7 m) would be possible, with the loss of half of Florida the result.

Since around three billion people – half the world's population – live near sea level, the impact of rising seas will be enormous. Farmlands will be flooded, shorelines eroded, coastal soils and water supplies contaminated by salt water, and mangrove forests and coral reefs destroyed. Coastal communities will be battered by higher storm surges and, because of warmer tropical seas, more frequent and more devastating hurricanes.

The Maldives nation in the Indian Ocean consists of 180,000 people living on 1,200 atolls. These islands would virtually disappear beneath the waves if sea level rose by 3 feet (1 m). A similar fate could befall Florida Keys, New Orleans, Shanghai and Venice. One-eighth of Bangladesh would be submerged leaving 10 million people homeless.

Major river deltas of the world including the Ganges–Brahmaputra, the Mekong, the Nile, and the Mississippi deltas would be threatened. The cost of protecting vulnerable coastlines worldwide will be astronomical. Developing nations have little hope of meeting these costs unless helped by developed nations.

The impact of future global warming and rising sea level cannot be predicted with certainty but the threat is so serious that nations are beginning to take some action, albeit slowly. Major reductions in greenhouse gases are needed and much preparation is required if vulnerable nations are to adapt to a rapidly changing climate. Coping with this global climate experiment unwittingly initiated by human activities provides a challenging test of the ability of nations to work together for the collective benefit of all people of the world, the next generation, and the environment.

Drastic cuts in carbon dioxide emissions are needed to reduce atmospheric levels of this greenhouse gas and thus limit global warming. Renewable energy sources such as wind, solar, hydroelectric, tidal, wave and geothermal energy, which do not emit carbon dioxide, should be substituted for some electricity generation that uses coal and oil.

California leads the world in harnessing the wind, using groups of wind turbines as seen here at Altamont Pass wind farm. Each turbine generates only a few watts of electricity, but there are 20,000 wind turbines operating in the state. A large single turbine is striking but clusters can appear intrusive, spoiling landscape views.

To overcome this problem and the need for large sites many wind farms are located on offshore breakwaters in Denmark, and Britain plans to place some in its offshore shallow seas. Several nations aim to produce 10–20 percent of their energy from wind farms by next century.

Re-creating on a small scale the Sun's phenomenal power output is the ultimate energy source. At incredible temperatures and crushing pressures in the Sun atoms are fused, releasing vast energy. By contrast, nuclear reactors use fission to produce energy by splitting atoms, but this uses radioactive materials which when spent have to be disposed of. After decades of attempts to generate sizable energy from fusion, scientists succeeded in 1991. Now the first large fusion reactor producing clean, almost inexhaustible, energy is promised within 25 years.

Changing worlds

Since the Earth formed more than 4,000 million years ago its atmosphere, oceans, continents and living organisms have experienced many profound changes. In recent decades human actions have increased the speed with which the world's environment is changing especially as world population grows by 100 million a year – three people every second.

The rapid consumption of energy and natural resources and the degradation of regional ecosystems is altering Earth's vital life-support cycles of energy, water, matter and chemicals. Yet natural events can initiate major and rapid changes in the world too. These have occurred many times during the Earth's lifetime and will continue to do so.

When Mount Pinatubo in the Philippines erupted in 1991 it ejected millions of tons of ash, dust and gases into the atmosphere. Large amounts of sulphur dioxide reached the stratosphere, where it combined with water to form minute droplets of sulphuric acid which girdled the Earth in a widening band. These tiny droplets scatter incoming sunlight, causing the Earth's surface to cool by as much as 0.9°F (0.5°C) for several years, temporarily counteracting the warming caused by the human-enhanced greenhouse effect.

But the ozone layer was damaged by the eruption since the sulphuric acid droplets provided surfaces on which chlorine could react with the ozone and hasten its destruction. Ozone losses of 10-15 percent allowed increased ultraviolet radiation to reach the surface damaging plants and phytoplankton and increasing the chances of skin cancer.

The world's climate is influenced by many factors operating on different time scales. Human activities are adding greenhouse gases to the atmosphere, forcing the planet to warm. Sulphates – acid rain pollutants – have increased greatly over industrialized regions, scattering sunlight and lessening global warming during the day.

One key to climate control is how much sunlight is reflected back into space. Replacing forests with fields and grasslands with deserts alters surface reflectivity, or albedo. Since agriculture developed, the planet's reflectivity has increased by 10 percent. Important reflectors of sunlight are clouds. Irrigation, aircraft exhausts and fossil fuel burning encourage cloud formation, whereas dry lands produced by overgrazing and soil erosion inhibit them.

Ice sheets, snow and sea ice also reflect much sunlight. Given a slight annual increase in area, a snowball effect can occur whereby more ice reflects more sunlight, causing cooling, which in turn encourages more ice, and

so on. The trigger for this can be changes in the Sun's energy reaching the planet, since this varies over thousands of years as the Earth undergoes cyclic changes in its orbit.

Ocean currents share out warmth around the world. If they alter in strength or direction climate changes. During the last ice age the Gulf Stream shifted south, flowing toward Portugal instead of Iceland and denying northern Europe vital warmth.

Volcanoes pour dust and gas into the stratosphere, forming a sunscreen, cooling the surface and

causing unseasonal weather as the jet stream is pushed equatorward. Eruptions millions of years ago caused months of darkness during which many species failed to survive. Such eruptions will happen again but there is no knowing when. Large meteors striking the Earth and throwing up huge dust clouds have similar effects.

Over millions of years climate and global weather systems alter greatly as the continents change latitude due to sea-floor spreading and the movement of tectonic plates.

Volcanic activity

Meteor impact

Rocky Mountains

Changes in
chemical balance
of atmosphere

Variations in
amount of solar
energy received

Changes in landforms

Human intervention
may disturb natural balance

Changes in cloud cover
and weather systems

Ice cover affects
albedo and sea level

Plate tectonics
alters size and
shape of
oceans and
continents

Jet stream

Circulation of winds and tides
moves energy around planet

Mid-Atlantic Ridge

195

Changing worlds/2

Heat transfer processes in the oceans are affected by volcanic eruptions. Regional cooling directly beneath the aerosol veil creates a temperature contrast with adjacent regions, generating winds which alter ocean currents. The eruptions of both El Chichon (1982) and Mount Pinatubo marked the start of El Niños, when abnormally warm waters spread across the central and eastern equatorial parts of the Pacific Ocean. El Niños add warmth to the atmosphere – even offsetting the cooling caused by the volcanic eruption – and cause drought in Australia, flooding in Peru and unseasonal weather elsewhere.

Volcanic eruptions a thousand times more powerful than modern eruptions occurred millions of years ago, bringing months of darkness and several years of twilight chill during which many animal and plant species became extinct. Life on Earth continued as new life forms evolved to take the place of those lost. In other words, even cataclysmic changes in world climate and environment were but temporary setbacks in the evolution of life. Is there a lesson to be learned from this?

Many people are increasingly concerned that polluting and degrading the environment, destroying the rich ecosystems of tropical rainforests and coral reefs, consuming natural resources, and triggering extinction of many species may be causing irreparable harm to our planet. However, given the problems that life on Earth has overcome – massive volcanic eruptions, meteor bombardments, a brightening sun and an initially inhospitable atmosphere – it seems likely that life in some form will survive even our worst excesses.

What is less certain is whether our own species will survive. If we trigger our own demise and another species replaces us it would not be that unusual, since, relatively speaking, few species last very long on our planet – five million years is the average. What appears unique about our species compared with others that have dominated the planet in the past is that we have the ability to recognize the global problems we are causing. Whether we really are the "wise, wise humans" that our full species name *Homo sapiens sapiens* suggests

and can avoid extinction will probably be discovered within the next few centuries.

Our actions do not yet appear to have pushed Gaia, our living planet, to the critical limits beyond which balance cannot be restored using its existing global regulating systems. If it had we would be experiencing runaway changes which could only be stabilized by Gaia readjusting its cycles to a different type of world, for example a much hotter or drier one. How long we have before such a change is effected is not known but it seems we have a little time in which to take responsibility for ensuring that the Earth's present global balance is maintained.

This requires a radical rethink of many of our activities. Indiscriminate exploitation, destruction and waste proliferation need to be replaced by sustained development, conservation and waste recycling. After all, the Earth does not need humans to survive but we need the Earth.

An increase in the Sun's activity is revealed by this gigantic twisted loop of erupting gas. Such activity creates a sudden surge in the electrically charged particles in the solar wind which stream toward the Earth, resulting in spectacular shimmering aurorae across the subpolar night skies.

Such displays indicate how responsive our planet can be to changes in solar activity. When the Sun's energy output fell by less than 1 percent between 1450 and 1850 the result was the Little Ice Age, a time of bitterly cold winters, dismal summers, failed harvests and violent storms.

Clearly the Sun affects world climate, but despite an increase of 30 percent in the Sun's energy since the Earth was formed, global average temperature has remained within life-sustaining limits. Even the

ice ages left the tropical regions largely unaffected.

As life evolved, by virtue of its existence, it gradually modified its environment to ensure that the climate remained within tolerable limits. As this process continued, conditions suitable for the development of humans were eventually created.

Humans, too, are now changing the environment, but these changes may be taking place too fast for the planet to redress the balance, especially as we are damaging its restorative mechanisms, for example the rainforest and the ozone layer.

If we want to continue as at present, we must now take responsibility for ensuring that the favourable planetary conditions which allowed us to evolve from apes 5–10 million years ago are not radically altered because of our actions.

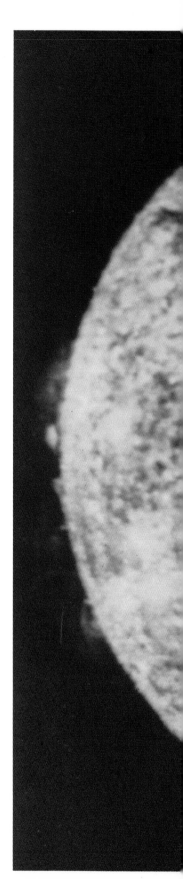

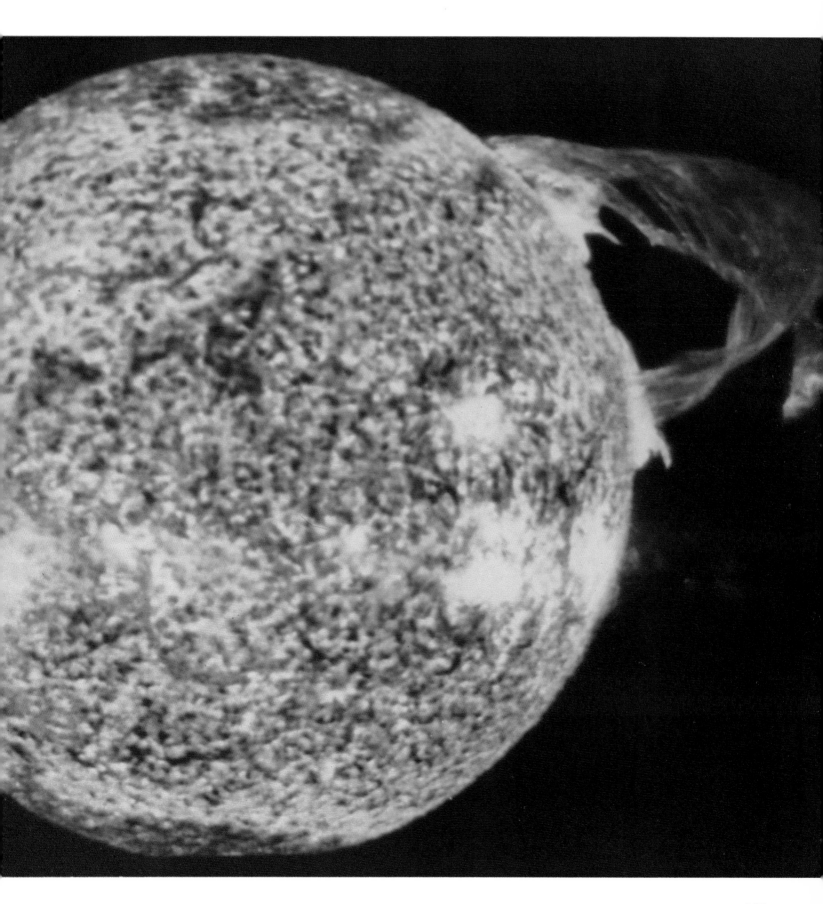

Glossary

Words in **bold** indicate that there is a separate entry under this subject. Words in *italic* are as important as the headword. The glossary should be used in conjunction with the index, since some subjects are treated in the text as well.

A

ablation
The loss of ice and snow from a glacier, whether by melting, **evaporation** or breaking away.

abrasion
The wearing away of part of the Earth's surface, caused by sand or other hard materials being ground against the surface by the action of wind, water or glacial ice.

absorption
The process by which a substance retains a gas, liquid or radiant energy. Radiant energy – such as that provided by the Sun's rays – is taken up and converted into heat or another form of energy.

abyssal plain
A flat region of the ocean floor which lies deeper than the **continental shelf** but not as deep as the **ocean trenches** i.e. between about 3,300 and 33,000 ft (1,000 and 10,000 m). At this depth there is virtually no light, so the abyssal plain supports very little life and the temperature rarely rises above 39°F (4°C).

acid rain
Rainwater with an abnormally acidic pH balance (between 2 and 5, which is 100,000 to 100 times more acidic than distilled water). The increase in acidity is due to the presence of large quantities of sulphur- and nitrogen-bearing gases, caused by burning oil, gas and coal. Acid rain is harmful to soil and plants and can kill water-dwelling plants and animals. It also corrodes stonework.

acid rock
An **igneous rock** that is more than two-thirds silica, often occurring as quartz or granite.

aftershock
A smaller **earthquake** following the main tremor.

air mass
A body of air in the lower **atmosphere** which is more or less constant in temperature and moisture content and may be bounded by **fronts**.

albedo
The ratio of solar radiation reflected by a surface to that received by it, usually expressed as a decimal or percentage. The Earth's average albedo is 0.3 (30%).

alluvium
A combination of silt, sand and gravel carried in suspension by a river and deposited on the riverbed or on the **flood plain** or **delta**. The deposits of **sediment** as a river enters a plain or open valley form an *alluvial fan* and may join other fans created by nearby rivers to form an *alluvial plain*.

Antarctic zone
The area defined approximately by the lines of latitude 60° and 75°S and including the Antarctic Circle (66$\frac{1}{2}$°S). The subantarctic zone lies to the north and the **polar zone** to the south.

anticline *see* fold

aquifer
A layer of rock such as chalk or sandstone which can store a large amount of **groundwater**, but also allows the water to permeate through it.

Arctic zone
The area defined approximately by the lines of latitude 60° and 75°N, including the Arctic Circle (66$\frac{1}{2}$°N) and corresponding to the **Antarctic zone** in the southern hemisphere.

arête
A narrow ridge between two glacial valleys. The characteristic jaggedness is caused as the **glaciers** erode the ridge from either side.

artesian well
A well in which the water is forced upward by hydraulic pressure, which builds up because the outlet to the well is below the **water table** and the water is held in a confined space between impermeable layers of rock.

asthenosphere
The part of the Earth's **mantle** directly below the **lithosphere**, about 60-220 miles (100-350 km) below the surface, where the rock is weak and easily deformed. It is because of movements in the asthenosphere that **plate tectonics** occurs.

atmosphere
The envelope of gases surrounding the Earth, which are held in place by the force of gravity and become less dense the farther they are from the Earth.

atmospheric pressure
The pressure exerted by the atmosphere because of the effect of gravity on the air above the Earth's surface. Like the force of gravity, atmospheric pressure decreases with altitude.

atoll
A ring of coral reef which may be broken or continuous, but always encloses an area of water. Especially in the Pacific Ocean, archipelagos are often made up of a number of atolls.

aurora
A display of changing coloured light in the night sky, usually near one of the magnetic poles, where rapidly moving charged particles from the Sun are trapped in the Earth's magnetic field.

B

barrier island
A long, narrow island, usually sand, parallel to the mainland and separated from it by a lagoon. A coral reef in a similar position is called a *barrier reef*.

basalt
A fine-grained **basic rock**, usually occurring in lava flows and comprising over 90% of volcanic rocks. *See also* flood basalts.

basic rock
An **igneous rock** made up of about 45-55% silica, i.e. appreciably less than **acid rock**. Basic rocks are often green or black in colour because of the other minerals which they contain.

batholith
A large mass of **igneous rock** formed when an eruption of **magma** spreads through the existing rocks underground. It extends to unknown depths below the surface. Batholiths form the substructure of many mountain ranges and are revealed when softer rocks above them are worn away.

bed load
The **sediment** rolled and bounced along the riverbed. During exceptional floods large boulders may form part of the bed load.

bedding plane
A surface separating two layers of **sedimentary rock** which have been formed at different times and have distinct characteristics.

bedrock
On land, the continuous layer of solid rock which constitutes the top of the Earth's crust. It may have no **soil** cover, but more often lies beneath a variable depth of soil, subsoil and loose material or **regolith**.

Benioff zone
A seismically active zone stretching downward from an **ocean trench** between the Earth's crust and the **asthenosphere**.

biomass
The total weight of living matter – or of a particular species – within a particular area or **ecosystem**.

biome
A geographical region – for example desert, grassland, forest – on land or sea, a subdivision of an **ecosystem**, characterized by a certain sort of vegetation and the animal life associated with it.

bluff
An almost perpendicular cliff face, usually along a river valley and formed by the action of the river. *See also* meander.

brickfielder
In Australia, a hot, dry, summer wind, particularly common in Victoria. It blows southeastward from the interior, bringing large quantities of dust and is accompanied by extremely high temperatures.

butte
A steep-sided, flat-topped hill or peak often found in the western United States. The top layer of a butte is hard rock which resists **erosion** and prevents the softer rock below it being worn away. *See also* mesa.

C

caldera
A deep, round basin formed after a violent volcanic eruption when the **volcano** collapses into its own **magma** chamber. If the volcano is extinct, the caldera will usually be filled with water forming a lake; if not, it may contain one or more potentially active secondary cones.

canyon
A large but comparatively narrow, steep-sided river valley formed by the river cutting through rock. Canyons are most common in arid regions because the lack of rainfall minimizes **erosion** and thus keeps the walls steep.

carbonates
A group of minerals which are important components of a type of **sedimentary rock** known as carbonate rocks. The most commonly occurring is calcium carbonate found in **limestone**, but carbonates of magnesium, manganese, iron and other metals are also found.

channel
The bed and sides of a watercourse, usually deep and narrow. The term may apply to a river, to a stretch of water connecting two larger areas of ocean or to a less clearly defined navigable part of a waterway.

chemical weathering *see* weathering

cirque
The source of a valley **glacier** – a steep-sided, circular indent in a mountainside, caused by glacial action or frost. Where the glacier has melted, the cirque may contain a lake whose sides are defined by moraine, or glacial debris.

climate
The pattern of weather in a given area averaged out over a long period of time. **Latitude**, altitude, prevailing winds and **ocean currents** all play a part in determining climate, although **precipitation** and temperature are the most important variables.

climax
The final stage in the process of **ecological succession**, when a stable community of plants and animals becomes established.

clint
A small segment of a limestone pavement or bedding plane, bordered by steep, narrow fissures, or **grykes**.

cloud
A mass of water droplets or ice crystals (or both) which is formed when water vapour condenses in the **atmosphere**. Many millions of these tiny suspended droplets or crystals have to combine to form droplets heavy enough to fall as **precipitation**. Clouds are classified according to their appearance and the height at which they appear.

coastal plain
The low-lying area between the ocean and the nearest high land. It may be formed by deposits of **sediment**, by a fall in sea levels or by rising land.

cold front *see* front

condensation
Process by which water vapour changes to liquid water.

continental shelf
The area of the ocean floor closest to land. It slopes gently from the shore line until it reaches a depth of about 425 ft (130 m), when the ocean floor falls away steeply toward the **abyssal plain**. The higher part of the steep slope is called the *continental slope*; as it approaches the abyssal plain the slope moderates to a slight gradient and becomes the *continental rise*.

convection current
A current caused by thermal convection, the process by which a warm gas or liquid rises and is replaced by a cooler, denser gas or liquid. Convection currents are found in the **atmosphere**, in the sea and within the **asthenosphere**. A strong individual circulation current is known as a *convection cell*.

coral reef
A ridge of rock produced largely by the accumulation of vast numbers of skeletons of stony (true) corals mingled with sand and shell fragments.

core
The centre of the Earth, the outer boundary of which is marked by the **Gutenberg discontinuity**. The *outer core* seems to be a dense liquid nickel-iron mixture, while the *inner core* is even more dense and may be solid.

Coriolis force
An effect produced by the Earth's rotation which, in the northern hemisphere, causes the wind (or any object moving relative to the surface of the Earth) to deviate to the right of its expected course. In the southern hemisphere, the deviation is to the left. The force is at its maximum at the poles and negligible at the equator.

corrosion
Chemical **erosion** of a rock surface, particularly a riverbed, due to the interaction of salts dissolved in the water and minerals contained in the rocks. *See also* weathering.

crater
The hollow at the top of a **volcano** which leads down into the channel or pipe from which **magma** emerges in the course of an eruption. In an extinct volcano, water may accumulate in the crater to form a *crater lake*. A crater may also be formed by the impact of a **meteor** striking the Earth's surface.

crevasse
A deep crack in the surface of a **glacier** which occurs when an alteration in the terrain such as a steep slope or a sharp turn has disrupted the glacier's movement.

crevasse splay
A fan-shaped area of **sediment** laid down when the waters of a rising river break through a **levee** on to the **flood plain** beyond.

crust
The thin outer shell of the Earth, a solid layer composed largely of silicates. It is about 12-25 miles (20-40 km) thick on landmasses (*continental crust*) and 4-9 miles (6-15 km) thick beneath the oceans (*oceanic crust*). The lower boundary is marked by the **Mohorovicic discontinuity** or moho.

cyclone
A system of low **atmospheric pressure** in which surface winds blow inward – counterclockwise in the northern hemisphere and clockwise in the southern – and ascend, thus leading to cloudy skies. Cyclones formed in the middle latitudes are often called **frontal depressions**. Those formed in low latitudes are called tropical cyclones (referred to as *hurricanes*, cyclones or *typhoons* in different parts of the world).

D

deflation
The lowering of a land surface caused by loose, dry particles of soil being carried away by the wind. The *deflation hollow* so caused may descend as low as the **water table**, at which point the soil is too damp for the wind to carry it and a lake or **oasis** may form.

deforestation
The felling and clearing of forest by human activity.

degradation
Generally, the lowering of the land surface by physical processes; specifically, the **erosion** of the **alluvium** or **bedrock** of a riverbed, resulting in the deepening of the channel.

delta
The alluvial plain (*see* alluvium) at a river mouth, often very fertile. It is caused by a build-up of **sediment** which is deposited more quickly than it can be carried away by the sea or lake currents; as it washes against these accumulations of sediment, the river is forced to subdivide into smaller channels which may spread out into the typical fan shape.

denudation
The result of severe **erosion** of the land by physical or chemical factors, and one of the major causes of changes in the Earth's surface. *See also* deposition.

deposition
The laying down and accumulation of solid particles carried by wind or water from one place to another. One of the two factors in the sculpting of the Earth's surface, the other being **denudation**.

depression *see* frontal depression

desert
An area where the rainfall is so little or so irregular that very little life can survive, though it is rare for there to be no vegetation at all. Anywhere that receives less than 10 inches (250 mm) of rain a year is normally considered a desert.

desertification
The spread or development of desert, characterized by worsening soil quality and plant cover, in an area that could once have supported crops.

doldrums
A zone of low pressure near the equator, lying between the trade wind zones, characterized by light and variable winds.

drainage basin
The area occupied by a *drainage system*, comprising all the smaller streams and watercourses which feed into a larger body of water, and the land from which water may flow. The highest point in a drainage basin, marking the boundary between one drainage system and the next, is called the *watershed*.

drought
A long period without rain, or with appreciably less rain than average.

drumlin
A ridge of glacial deposits and debris occurring in a once glaciated area. Drumlins usually occur in groups known as a *drumlin swarm*. Its length runs parallel to the direction of the flow of the **glacier**.

dune
A mound of sand deposited by the wind, found in a desert or near the seashore. The action of the wind continues to lift the sand so that it falls over the peak of the dune on to the leeward side; dunes are therefore constantly changing in shape and gradually moving in the direction of the prevailing wind.

dyke
A sheet of **igneous rock** formed by **magma** that rose from a magma chamber and cut through the surrounding strata.

E

earthquake
A trembling of the ground usually caused by **seismic waves** and most frequent along a **fault**. The size of an earthquake – measured on the **Richter scale** – depends on the amount of energy released when the brittle rocks of the fault fracture. The areas most prone to earthquakes are those at plate boundaries.

ecological succession
The chronological sequence of plant and animal communities that colonize a given area of newly formed or newly cleared land. *See also* climax.

ecosystem
An ecological system, a community of living organisms in their habitat. It may be clearly delineated, like a lake or pond, or loosely defined and merge with neighbouring ecosystems. The living creatures within each ecosystem form an interdependent food chain. *See also* biome.

El Niño
A reversal of prevailing trade winds and ocean currents across the Pacific Ocean, which happens every few years with catastrophic effects on weather in Australia and parts of South America.

epicentre
The point on the Earth's surface directly above the **focus** of an **earthquake**, where the earthquake is at its most intense; most damage is caused here.

equinox
Either of the two points in the calendar at which the path of the Sun intersects with the plane of the equator, making day and night of equal length. In the northern hemisphere the *spring or vernal equinox* occurs about March 21 and the *autumn equinox* about September 21. At noon on these days the Sun is directly overhead at the equator.

erosion
The process by which the surface of the Earth is worn away by the action of water, ice, wind and other physical factors; chemical erosion may occur because of the interaction of salts dissolved in water with minerals found in rocks. *See also* weathering.

esker
An elongated ridge of glacial deposits, sometimes hundreds of miles long. It consists chiefly of sand and gravel left once the ice has melted or retreated.

estuary
The widening, tidal part of a river near its mouth, where fresh and salt waters mix.

evaporation
The process by which any liquid is converted into gas; specifically, the process by which water is absorbed into the atmosphere, an important element in the **water cycle**.

evapotranspiration
Water lost to the atmosphere from soil and plants through **evaporation** and *transpiration* respectively.

F

fault
A weak line or zone in the rocks of the Earth's crust, where the rocks crack and move in relation to each other on either side of the weakness, causing an **earthquake**. The change in position may be fractions of an inch or many hundreds of yards.

fjord
A long, deep and narrow inlet from the ocean, bounded by steep-sided cliffs, carved by glacial action and especially common on the west coast of Norway.

fissure
An extensive crack in a rock. In volcanic areas, **lava** may be expelled through a fissure without an explosion (*fissure eruption*). *See also* flood basalts.

flood basalts
Thick layers of **basalt** spread over large areas as a result of volcanic outpourings.

flood plain
The low-lying area on either side of a river or other watercourse over which the water floods. The soils of the flood plain are largely composed of **alluvium** left behind when the flood waters retreat.

focus
The point beneath the Earth's surface where an **earthquake** starts, i.e. the point directly below the **epicentre**. It is usually found less than 45 miles (70 km) below the surface (*shallow focus*), but may occur as deep as 185 miles (300 km) (*deep focus*).

fold
The bending or buckling of layered or **sedimentary rock** as a result of compression. A *syncline* is a U-shaped fold in the **strata** in which the youngest rocks are in the centre; in an *anticline* the strata are deformed upward with the oldest rocks on the inside.

fossil
Remains of a plant or animal that has been buried for a long period. Most fossils are found in **sedimentary rock**, but they may also occur in volcanic ash or mud which preserve the dead organism from decay.

fossil fuel
A naturally occurring hydrocarbon fuel derived from the remains of organic material encased in rock; the most common examples are coal, petroleum (crude oil) and natural gas.

front
A narrow, sloping layer of air between two air masses of different temperature, associated with distinctive cloud and weather patterns and important in short-term weather forecasting. A cold front is the leading edge of a cold air mass; a warm front leads a warm air mass. When a cold air mass overtakes a warm one, forcing the warm air to rise and cool, a composite or *occluded front* is formed. *See also* occlusion.

frontal depression
An area of low pressure in the mid-latitudes, characterized by unsettled weather and rainfall. *See also* front, cyclone.

fumarole
A hole in the Earth's crust, common in volcanic regions, through which steam and gases such as carbon dioxide and sulphur dioxide are released under pressure.

G

geology
The study of the Earth, its origins, composition and the forces acting upon it.

geomorphology
The branch of geography concerned with landforms, their origins and development.

geothermal heat
The heat of the Earth's interior, producing *geothermal energy* which is used in some parts of the world, such as Iceland, to produce electricity.

geyser
A natural fountain of hot water or steam ejected periodically from a hole in the Earth's crust. Found in areas of volcanic activity, geysers are the result of underwater streams coming into contact with hot, volcanic rocks beneath the surface of the Earth.

glacier
A mass of ice originating on land, produced by the accumulation and compression of snow and often extending far below the snow line. It moves gradually downhill under gravity, following valley contours. Glaciers pick up rock material, sand and gravel as they go, which they deposit as moraine as the ice melts at lower and warmer altitudes.

Gondwana
The southern supercontinent which 180 million years ago comprised modern-day South America, Africa, Arabia, India, Australia and Antarctica. After that time, the continents began their gradual drift to their present positions. *See also* Laurasia, Pangaea.

graben *see* rift valley

granite
A coarse-grained **plutonic igneous rock** containing quartz, feldspars, mica and other minerals.

greenhouse effect
The accumulation of heat in the lower **atmosphere** caused by the retention of some infrared **radiation** emitted from Earth. Gases, especially carbon dioxide, CFCs and methane, allow sunlight to pass through but absorb infrared radiaton emitted by the Earth which would otherwise be transmitted to space.

gryke
A joint in a slab of **limestone** widened by **weathering** and often forming the border of a **clint**.

groundwater
Water found under the surface of the Earth, saturating pores and crevices. Most of it enters as rainwater into soil and **sedimentary rock**, where it contributes to **weathering** and chemical changes, dissolving certain rock materials and depositing dissolved minerals. The surface of groundwater, which varies according to the amount of rain that has fallen, is known as the *water table*. *See also* aquifer.

groyne
A man-made barrier running down a beach into the sea, designed to trap **sediment** carried along the coast by **longshore drift** and thus retain a beach.

Gutenberg discontinuity
The boundary between the Earth's solid mantle and liquid outer core. It occurs at a depth of 1,800 miles (2,900 km) and has a marked effect on **seismic waves**.

guyot
A volcanic, flat-topped, submarine hill carved by the action of waves.

H

hanging valley
A valley which enters another from a higher altitude, so that one river joins another by means of a waterfall or rapids. A hanging valley is usually created by a small **glacier**, a tributary of the main glacier which has carved the lower, larger valley.

I

ice age
A period in geological history characterized by global lowering of temperatures, when ice spread beyond the polar regions over large tracts of the Earth's surface. An ice age usually lasts from one to three million years; the last one began about two million years ago. Within ice ages there may be short cold or warm periods, called *glacials* and *interglacials* respectively, when the ice advances and retreats.

ice cap
An area of permanent ice usually, though not necessarily, smaller than an *ice sheet*, which is a thick plate of glacial ice covering a landmass. The ice sheet covering Antarctica contains 85% of the world's ice. An *ice shelf* is similar to and attached to an ice sheet, but lies over water rather than land.

igneous rock
One of the three main kinds of rock that make up the Earth's crust, the others being **metamorphic** and **sedimentary rocks**. Igneous rocks are formed when a hot, molten substance such as **magma** cools and solidifies. This may occur on the surface of the Earth; in a channel linking the molten material to the surface; or well below the surface, under pressure.

impermeable rock
A rock through which water cannot percolate.

infiltration
The absorption of **precipitation** and surface water into rocks and soil through cracks or porous material.

insolation
An abbreviation for <u>in</u>coming <u>sol</u>ar radi<u>ation</u>. The Sun emits about half of its energy in the form of visible radiation (light) and the rest as ultraviolet and near-infrared radiation.

interglacial *see* ice age

intrusive
Describes a mass of **igneous rock** formed when **magma** solidifies between layers of existing rock.

J

jet stream
A narrow area of strong winds which may reach speeds of 300 mph (480 km/h) in winter. Jet streams occur where sharp changes in temperature produce great differences in pressure over short distances.

joint
A crack in a solid piece of rock, differing from a **fault** in that there is no movement between the blocks.

K

karst
A **limestone** region with a dry surface, often of barren rock, and a network of underground channels and caverns, named after a region of Slovenia where these features occur.

kettle hole
A hollow in a land surface caused when a block of ice covered by gravel has melted, allowing the glacial deposits to settle. It often contains a lake.

L

lagoon
A body of water virtually cut off from the open sea by an **atoll**, by a narrow spit of land, by a barrier beach or **barrier island,** or by a reef.

landform
A distinctive feature of the land surface, such as a hill, valley or plateau, produced by physical processes.

latitude and longitude
The means by which any point on the Earth's surface may be defined. Lines of latitude encircle the Earth parallel to the equator, latitude 0°.

Lines of longitude pass through both poles and cross the equator at right angles. The Greenwich meridian is 0°, the International Date Line 180°; the areas in between are divided into degrees east and west of Greenwich, and around the globe time zones are calculated as being a number of hours behind or ahead of Greenwich.

Laurasia
The northern supercontinent which after the breakup of **Pangaea** comprised modern-day Greenland, North America, Europe and Asia, excluding India. *See also* Gondwana.

lava
The name given to **magma** when it emerges from a **volcano** and spreads over the Earth's surface, where it quickly cools and solidifies.

leaching
The continual action by which water percolating downward through the soil removes soluble material such as mineral salts.

levee
A natural river bank built up by deposits of **sediment** during flooding; it remains when the waters retreat and is increased by further deposits the next time the river floods. A levee is therefore the highest part of a **flood plain** and may be a large enough obstacle to divert or divide the course of a river. Levees are frequently raised and strengthened artificially to reduce the risk of flooding.

limestone
A **sedimentary rock** composed mainly of calcium carbonate. It is soluble in water containing carbon dioxide. Organic limestones, of which chalk is the most common form, are formed from **fossil** material. Limestone or **karst** regions may be characterized by underground caverns and drainage systems.

lithosphere
The solid and brittle outer layer of the Earth which includes the crust and part of the upper **mantle**. It may be up to 60 miles (100 km) thick in some places and is bounded on the inside by the **asthenosphere**.

longitude *see* latitude

longshore drift
Also known as beach drift. The movement of beach materials, such as sand and pebbles along a coast, due to waves breaking on the beach at an angle.

M

magma
The molten material below the solid rock of the Earth's crust. It is composed largely of silicate, with dissolved gases and sometimes suspended crystals. Magma is thrust out of a **volcano** during an eruption, when it solidifies to become **lava**. If it fails to reach the surface but cools underground, it forms **intrusive igneous rock**.

magnetosphere
The zone around the Earth occupied by the Earth's magnetic field. It is pushed into a teardrop shape by the solar wind, to extend much farther on the side away from the Sun.

mantle
The layer between the Earth's crust and the **core**. The mantle is about 1,783 miles (2,870 km) thick and consists of **igneous rocks** of silicate composition, of higher density than the rocks which make up the crust. The upper boundary with the crust is marked by the **Mohorovicic discontinuity** while its lower boundary with the outer core is marked by the **Gutenberg discontinuity**.

marine terrace *see* terrace

mass movement
Any of a number of processes, including landslides, mudflow and the more gradual soil creep, by which large quantities of material move down a hill, slope or cliff.

meander
A bend or loop in the course of a river that curves back and forth across a **flood plain**. The action of the current wears away the bank on the outside of a curve, forming a **bluff**, while depositing **sediment** on the inside, forming a **point bar**. This increases the size of the meander, which may bend around almost far enough to form a circle. When the water is high it may cut a new, direct path across the neck of the meander. Deposits of sediment gradually seal the meander off from the river resulting in a crescent-shaped lake called an *oxbow lake*.

mesa
From the Spanish word for table, a flat-topped hill or plateau falling away steeply on all sides. The top layer consists of hard rock that has resisted erosion, while softer surrounding rock has been worn away. It is normally larger than a **butte** and may, with further erosion, divide and form a number of buttes.

metamorphic rock
One of the three main types of rock that make up the Earth's crust, the others being **igneous** and **sedimentary rocks**. Metamorphic rocks were formerly either igneous or sedimentary but have been changed in character and appearance by the action of heat and pressure.

meteor
A streak of light in the night sky, caused by a fragment of rock or dust particle from space entering the Earth's atmosphere and igniting. Meteors are usually tiny and burn up in the atmosphere, forming *shooting stars*. However, larger ones occasionally fall to Earth, sometimes leaving large craters. The material thus deposited is known as a *meteorite*.

meteorology
The study of the **atmosphere** and its interactions with the ground surface. It is usually used to refer to the study of weather conditions in the troposphere and stratosphere.

mid-latitudes
The **temperate zones** lying between the **tropics** and the **Arctic** and **Antarctic zones**.

mid-oceanic ridge
A mountain range under the ocean, characterized by **earthquake** and volcanic activity. It marks a constructive plate boundary.

mistral
The cold, dry, northerly wind which blows down through the Rhône Valley in southern France toward the Mediterranean. It is caused by a depression (*see* frontal depression) in the Gulf of Genoa attracting cold air from the north.

Mohorovicic discontinuity (moho)
The boundary between the Earth's crust and **mantle**, characterized by a change in the velocity of **earthquake** waves and named after the Yugoslav seismologist who discovered it.

monsoon
A system of winds, occurring mainly in the **tropics**, whereby complete changes of wind direction from season to season bring extended periods of heavy rain and storm-force winds. The term is derived from the arabic word, *mausin*, meaning season.

moraine *see* glacier

O

oasis
Any place in the middle of a **desert** made fertile and habitable by the presence of water. *See also* deflation.

occlusion
The lifting of the warm front of a depression when a cold front catches up with it. A cold occlusion occurs when the air behind the cold front is colder than the air in front of the warm front; the reverse is called a warm occlusion. *See also* front, frontal depression.

ocean current
The persistent flow of ocean water in a given direction, determined by the prevailing winds and by some areas of water being more dense than others due to differences in temperature and salinity.

ocean trench
A deep, narrow depression in the ocean floor formed where the edge of one plate dives beneath another at a destructive plate margin, or *subduction zone*.

orographic rain
Rainfall over mountains, where the terrain forces air to rise, cool, condense and fall as precipitation.

outwash plain
The alluvial plain (*see* alluvium) built up when streams from a melting **glacier** carry moraine, usually sand and gravel, over a large, low-lying area.

oxbow lake *see* meander

ozone layer
Part of the layer known as the stratosphere in the **atmosphere**. It absorbs **ultraviolet radiation** from the Sun, preventing dangerously large quantities from reaching the Earth. Depletion of the ozone layer by pollutants, especially chlorine compounds, gives cause for concern, since greater amounts of ultraviolet radiation are increasing the risks of skin cancers and other diseases and causing damage to phytoplankton and zooplankton.

P

Pangaea
The hypothetical supercontinent that began to break up 180 million years ago when the continents gradually started to move to their present positions. Pangaea broke first into the two smaller landmasses, **Gondwana** and **Laurasia**.

peninsula
A strip of land surrounded on three sides by water.

permafrost
Permanently frozen ground, either soil or rock, in the **polar zones** extending to depths of 1,650 ft (500 m), although the surface may thaw in summer.

permeability
The degree to which rocks can permit water or other liquids to pass through them via pores in the grain of the rock. Sandstone is a commonly occurring *permeable rock. See also* pervious, porosity.

pervious
The term used to describe a rock which allows water to pass through it via joints and fissures, rather than pores. *See also* permeability, porosity.

photosynthesis
The process by which plants use carbon dioxide, water and energy provided by sunlight to produce the carbohydrates necessary for growth. The light is absorbed by the green matter known as chlorophyll, and oxygen is released as a by-product, replenishing the oxygen supplies in the **atmosphere**. This process is the starting point of an **ecosystem**.

plate tectonics
The theory that the Earth's crust, or **lithosphere**, is divided into relatively thin plates which move across the **asthenosphere** at the rate of $1/4$–$3^1/2$ in (1–9 cm) a year. The edges of the plates, the *plate margins*, are the areas prone to **earthquakes** and volcanic activity.

plucking
The lifting up and carrying away of quantities of **bedrock** beneath the ice as a **glacier** moves forward.

plutonic rock
A body of **igneous rock**, enclosed in existing rock, which has cooled slowly deep down inside the Earth. The slowness of the cooling process produces crystalline, coarse-grained rocks such as **granite**.

polar zones
The areas lying above latitudes 75° i.e. closest to the poles, characterized by **permafrost** and pack ice.

porosity
The degree to which rocks can hold water in pores or spaces between the grains of minerals. Porous rocks are not necessarily permeable (*see* permeability).

precipitation
Deposits of water from the **atmosphere** which reach the Earth in any form. These include dew and frost as well as rain, hail, sleet and snow. *See also* clouds.

pressure gradient
The measure of the difference in **atmospheric pressure** between two places. Winds blow from high to low pressure zones so the larger the pressure gradient, the stronger the winds.

pyroclastic material
Fragments of rock produced in volcanic eruptions, including volcanic ash and spindle bombs.

R

radiation
The transmitting of energy by electromagnetic waves. The wavelength of this radiation varies inversely with the temperature of the body or substance emitting the radiation. The intensely hot Sun emits relatively short-wave radiation, including light, while the cooler Earth emits long-wave radiation in the infrared bands.

rain shadow
An arid area on the leeside of a mountain range, created when moist air has been forced upward by the mountains, causing it to shed its moisture – known as **orographic rain** – on the windward side.

regolith
The layer of unconsolidated weathered material lying above the bedrock. The topmost layer is the soil.

Richter scale
The logarithmic scale which measures the size of an **earthquake**. The scale runs from 0-9, with the higher numbers designating the more violent quake.

rift valley
A valley formed by the sinking of a piece of land between two roughly parallel **faults**. The trenchlike block of sunken rock is called a *graben*.

river terrace *see* terrace

rock cycle
The cycle by which rock is created, changes character and is destroyed as a result of chemical and physical processes inside the Earth or on or above its surface. *See also* igneous, sedimentary and metamorphic rocks and weathering.

runoff
Water that flows across the land surface, often to a river or stream, instead of seeping into the soil.

S

saturated zone
The area beneath the Earth's surface where all the pores in the rock are filled with **groundwater**.

savanna
The tropical grasslands bordering the equatorial forests where annual dry seasons prevent the growth of more than the occasional tree.

scree *see* talus

sediment
Any solid material carried away from its point of origin and deposited elsewhere by water, ice or wind.

sedimentary rock
One of the three main types of rock that make up the Earth's crust, the others being **igneous** and **metamorphic rocks**. Sedimentary rocks are formed by an accumulation of **sediment** eroded from pre-existing rocks and deposited in layers – which are subsequently compressed together – by the action of wind, ice or water. *See also* weathering, limestone.

seismic wave
A shock wave emitted from the focus of an **earthquake** which causes the characteristic shaking.

seismology
The study of **earthquakes**.

shield
An ancient, large and stable area of the Earth's crust, at the centre of a continental plate. Shields are unaffected by violent plate activity; they are susceptible only to slow movement or to **erosion**.

sill
A sheet of **intrusive igneous rock** between two layers of **sedimentary rock**, caused by **magma** forcing its way between the layers and solidifying.

sirocco
The hot, dry, dusty, southerly wind that blows across Algeria and the Levant in spring and autumn and is drawn from north Africa toward the Mediterranean by **frontal depressions**. Moisture picked up over the sea causes cloudy conditions in southern Italy.

solstice
Either of the two points in the calendar at which the Sun is at its greatest angular distance north or south of the equator, giving, in the northern hemisphere, the shortest day about December 21 (*winter solstice*) and the longest day about June 21 (*summer solstice*). The situation is reversed in the southern hemisphere.

storm surge
An abnormal and rapid rise in the level of the ocean along a coast, principally as a result of high winds.

stratification
The layering of rock to form strata (*see* stratum).

stratum
A layer of rock, usually **sedimentary** but the term may apply to **igneous rock**. There is usually a series of parallel layers (*strata*), all more or less distinct and possibly with bodies of **intrusive** rock between them.

subduction zone *see* ocean trench

syncline *see* fold

T

taiga
A type of woodland found immediately south of the **tundra** zone, consisting of low-growing, widely spaced coniferous trees and a groundcover of lichens and mosses.

talus or scree
An accumulation at the bottom of a cliff or mountain slope of rock and rock fragments which have been broken from the main rock face by **weathering**.

temperate zone
The broad mid-latitude zone of mild but variable climate lying between the warm **tropical zone** and cold **polar zone**.

terrace
Part of a former **flood plain** raised above the level of the river **channel**. It may be created by water cutting the channel downward because of a change in the pattern of the river's flow; by a rising of the land; or by a lowering of sea level (*river terrace*). Also (*marine terrace*) a shelf, formerly under shallow water, now a dry coastal area, created by a lowering of sea level or a rising of the land.

terrane
An area of rock on the Earth's crust with a particular geological history and characteristics that differentiate it from the surrounding rock.

thermocline
The layer of ocean water, about 330–660 ft (100–200 m) below the surface, in which there are rapid changes in temperature.

tidal range
The average difference in water level between high and low tide at any given point. Around the time of the full and new moon, *spring tides* occur: these have a much larger than average range, with the high tides high and the low tides low. The opposite, *neap tides*, occur during the first and last quarters and have a much smaller than average tidal range.

tornado
A narrow violently rotating column of air extending from the base of a cloud, usually a thunderstorm, in the form of a funnel- or rope-shaped cloud. Its winds can reach 300 mph (480 km/h). A tornado forming on or travelling over water is called a waterspout.

trade winds
Winds, remarkably constant in speed and direction, which blow toward the equator in the subtropical belts, from the northeast in the northern hemisphere and the southeast in the southern hemisphere.

transform fault
A ridge or scarp between two plates on the ocean floor, at right angles to the **mid-oceanic ridge**.

tropical cyclone *see* cyclone

tropical zone
Loosely, the zone between the equator and the tropics of Cancer (23½°N) and Capricorn (23½°S) where the Sun is directly overhead for part of the year. More strictly, the *equatorial zone* is the area within 8° of the equator on either side, characterized by heavy rainfall and high temperatures, while the *tropical zone* begins where the equatorial zone ends.

tsunami
A tidal wave formed by an underwater **earthquake**.

tundra
The **Arctic zone** found north of the **taiga**, where much of the soil is subject to **permafrost**. Some vegetation, principally grasses, dwarf shrubs, mosses and lichens, grow during the brief summer.

U

ultraviolet radiation
Electromagnetic radiation beyond the violet (short-wave) end of the spectrum. The Sun produces large quantities of ultraviolet radiation, much of which is absorbed by the **ozone layer**.

unsaturated zone
The zone above the **saturated zone**, in which under normal circumstances the pores of the rocks are not completely saturated.

V

Van Allen belts
Two regions surrounding the Earth in which high-energy particles from the Sun are trapped by the Earth's magnetic field.

volcano
Strictly, a vent in the Earth's crust caused by **magma** forcing its way to the surface and through which molten or solid igneous material and gases erupt. However, the term is normally used for the cone-shaped mountain built up by the continuous deposits of **lava** and other debris emitted during an eruption.

W

warm front *see* front

water cycle
The cycle by which water changes state in a variety of processes including **precipitation, evaporation** and water flow. Water may fall as rain, run off the land into a river, be returned to the ocean and evaporate into the atmosphere, from which it will eventually fall again as rain.

watershed *see* drainage basin

water table *see* groundwater

weathering
The effect of decay or disintegration that exposure to the **atmosphere** has on the rocks of the Earth's crust. Weathering may be physical, the result of the action of rain, wind, heat or frost; or chemical, as when rainwater containing carbon dioxide dissolves **limestone** rocks.

Bibliography

Allen, O.E. *Planet Earth: Atmosphere* Time-Life Books, Amsterdam, 1983

Barry, R.G. & R.J. Chorley *Atmosphere, Weather and Climate* Methuen, London, 5th ed., 1987

Battan, Louis J. *Fundamentals of Meteorology* Prentice-Hall, Englewood Cliffs, New Jersey, 1979

Bloom, Arthur L. *The Surface of the Earth* Prentice-Hall International, London, 1979

Bramwell, Martyn (ed.) *The Mitchell Beazley Book of the Oceans* Mitchell Beazley, London, 1977

Brown, B. & L. Morgan *The Miracle Planet* Merehurst Press, London, 1990

Calder, N. *Spaceship Earth* Viking, Penguin Books, London, 1991

Calder, N. & J. Newell *Future Earth* Christopher Helm, London, 1988

Carpenter, Clive (ed.) *The Guinness Book of Answers* Guinness Publishing, 8th ed., 1991

Chorlton, W. *Planet Earth: Ice Ages* Time-Life Books, Amsterdam, 1983

Cocks, L.R.M. *The Evolving Earth* Cambridge University Press, Cambridge, 1981

Collins, M. (ed.) *The Last Rain Forests* Mitchell Beazley, London, 1990

Cattermole, P. & P. Moore *The Story of the Earth* Cambridge University Press, Cambridge, 1985

Cox, Barry & Peter D. Moore *Biogeography: An Ecological and Evolutionary Approach* Blackwell Scientific Publications, Oxford, 4th ed., 1985

Dixon, Dougal, Barry Cox, R.J.G. Savage & Brian Gardiner *The Macmillan Illustrated Encyclopedia of Dinosaurs and Prehistoric Animals* Macmillan Publishing, New York and London, 1988

Dixon, D. (ed.) *The Planet Earth: Seas, Climate, Continents* Leisure Circle, London, 1984

Dunning, F.W., P.J. Adams, J.C. Thackray, S. van Rose, I.F. Mercer & R.H. Roberts *The Story of Earth* HMSO, 2nd ed., London, 1981

Dury, George *An Introduction to Environmental Systems* Heinemann Educational Books, London and Exeter, New Hampshire, 1981

Durrell, Lee *State of the Ark* Doubleday, New York, 1986

Duxbury, A.C. & A.B. Duxbury *An Introduction to the World's Oceans,* Wm. C. Brown Publishers, Dubuque, Iowa, 3rd ed., 1991

Elsom, D.M. *Atmospheric Pollution: A Global Problem* Blackwell Publishers, Oxford U.K. and Cambridge U.S., 2nd ed., 1992

Farrand, J. Jr. *Weather* Stewart, Tabori & Chang, New York, 1990

Friday, Adrian & David S. Ingram (eds.) *The Cambridge Encyclopedia of Life Sciences* Cambridge University Press, Cambridge and New York, 1985

Fuchs, Sir Vivian (ed.) *The Physical World.* Oxford Illustrated Encyclopedia, edited by Harry Judge, vol. 1, Oxford University Press, Oxford and New York, 1985

Goldsmith, E. & N. Hildyard (eds.) *The Earth Report 2* Mitchell Beazley, London, 1990

Goudie, Andrew *The Human Impact on the Natural Environment* Basil Blackwell Publishers, Oxford U.K. and Cambridge U.S., 1990

—— *The Nature of the Environment: An Advanced Physical Geography* Basil Blackwell Publishers, Oxford U.K. and Cambridge U.S., 1984

Gregory, K.J. (ed.) *The Guinness Guide to the Restless Earth* Guinness Publishing, London, 1991

Gribbin, J. *Hothouse Earth: The Greenhouse Effect and Gaia* Bantam Press, Transworld Publishers, London and New York, 1990

Gribbin, J. & M. Kelly *Winds of Change: Living in the Global Greenhouse* Hodder & Stoughton, London, 1989

Hardy, R., P. Wright, J. Gribbin, & J. Kington *The Weather Book* Michael Joseph, London, 1982

Jackson, D.D. *Planet Earth: Underground Worlds* Time-Life Books, Amsterdam, 1982

Keller, E.A. *Environmental Geology* Merrill Publishing Company, Columbus, Ohio, 5th ed., 1988

Lambert, David & the Diagram Group *The Cambridge Guide to The Earth* Cambridge University Press, Cambridge, New York and Sydney, 1988

Lewis, T. (ed.) *Planet Earth: Volcano* Time-Life Books, Amsterdam, 1982

Lutgens, F.K. & E. J. Tarbuck *Essentials of Geology* Merrill Publishing Company, Columbus, Ohio, 1986

Marshall, Bruce (ed.) *The Real World* Houghton Mifflin Company, Boston and London, 1991

Myers, N. (ed.) *The Gaia Atlas of Planet Management* Pan Books, London and Sydney, 1985

Natural Wonders of the World Readers Digest, Pleasantville, New York and London, 1980

Pellant Chris (ed.) *Earthscope* Salamander Books, London, 1985

Pellant Chris *Rocks, Minerals & Fossils of the World* Pan Books, London, 1990

Redfern, Martin *Journey to the Centre of the Earth* Broadside Books, London, 1991

Ronan, Colin. A. *The Natural History of the Universe* Doubleday, London, New York and Sydney, 1991

—— *The Skywatcher's Handbook* Crown Publishers, New York, 1985

van Rose, Susanna *Earthquakes* HMSO for the Institute of Geological Studies, London, 1983

Skinner, Brian J. & Stephen C. Porter *Physical Geology* John Wiley & Sons, London and New York, 1987

Simmon, I. G. *Changing the Face of the Earth* Basil Blackwell Publishers, Oxford U.K. and Cambridge U.S., 1989

Smith, David G. (ed.) *The Cambridge Encyclopedia of Earth Sciences* Cambridge University Press, Cambridge, New York and Sydney, 1981

Smith, K. *Environmental Hazards* Routledge, London and New York, 1992

Smith, Peter J. (ed.) *Hutchinson Encyclopedia of the Earth* Hutchinson, London, 1986

Strahler, Arthur N. & Alan H. Strahler *Elements of Physical Geography* John Wiley & Sons, New York and Chichester, 4th ed., 1986

—— *Modern Physical Geography* John Wiley & Sons, New York and Chichester, 1992

Walker, B. *Planet Earth: Earthquake* Time-Life Books, Amsterdam, 1982

Whipple, A.B.C. *Planet Earth: Storm* Time-Life Books, Amsterdam, 1982

White, I.D., D.N. Harrison & S.J. Mottershead *Environmental Systems: An Introductory Text* Allen & Unwin, London, Boston and Sydney, 1984

Whitfield, Philip, Peter Moore & Barry Cox *The Atlas of the Living World* Weidenfeld & Nicolson, London, 1989

Whitfield, Philip (ed.) *Our Mysterious Planet: Mysteries of the Natural World* Cassell, London, 1991

Whittow, J. *Disasters: The Anatomy of Environmental Disasters* Allen Lane, Penguin Books, London, 1980

—— *The Penguin Dictionary of Physical Geography* Penguin Books, London, 1984

Wood, Robert Muir *Earthquakes and Volcanoes* Mitchell Beazley, London, 1986

Woodward, Christine & Roger Harding *Gemstones* British Museum (Natural History), London, 1987

Index

Acknowledgments

Contributors
Keith Addison
David Burnie
Jon Kirkwood
Helen Rudkin
Caroline Taggart

Project editor
Nigel Bradley
Art editor
Marnie Searchwell
Art assistants
Eileen Batterberry
Kate Harkness
Assistant editor
Lindsay McTeague
**Editorial research
& co-ordination**
Heather Magrill
Picture editor
Zilda Tandy
Indexer
Dorothy Groves

Editorial director
Ruth Binney

Production
Barry Baker
Janice Storr
Nikki Ingram

*The publishers would like to
thank Captain and Mrs
Moland for the use of their
daughter Sarah's books.*

Picture credits
l = left; *r* = right; *t* = top; *c* = centre; *b* = bottom

1 Jeff Foott/Survival Anglia; 2 Dorian Weisel/Planet Earth Pictures; 4*l* Bullaty/The Image Bank; 4*r* William Strode/ Susan Griggs Agency; 5*l* Jeff Foott/Oxford Scientific Films; 5*r* Larry J. Pierce/The Image Bank; 6/7 Kim Westerskov/Oxford Scientific Films; 8*l* Johnny Johnson/ Bruce Coleman; 8*r* N. Callow/NHPA; 9*l* Gunter Ziesler/ Bruce Coleman; 9*r* Michael Nichols/Magnum Photos; 13 Johnny Johnson/Bruce Coleman; 21 N. Callow/ NHPA; 23 Michael Nichols/Magnum Photos; 25 Peter Menzel/Science Photo Library; 27 Gunter Ziesler/Bruce Coleman; 28/29 & 30*l* Krafft/Explorer/Robert Harding Picture Library; 30*r* David Parker/Science Photo Library; 31*l* David Jeffrey/The Image Bank; 31*r* Brian Coope/ Planet Earth Pictures; 35 John Shaw/Bruce Coleman; 39 David Parker/Science Photo Library; 42 P. Vauthey/ Sygma; 42/43 John Lythgoe/Planet Earth Pictures; 45*t* David Paterson; 45*b* Eric Crichton/Bruce Coleman; 47 C. Weaver/Ardea; 50 Michael Nichols/Magnum Photos; 51 Y. Shone/Gamma/Frank Spooner Pictures; 54 Krafft/ Explorer/Robert Harding Picture Library; 55 R. Perry & J. Mason/Black Star/Colorific!; 56 Martyn F. Chillmaid/ Oxford Scientific Films; 57 A.N.T./Otto Rogge/NHPA; 58/59 Peter Menzel/Science Photo Library; 60/61 Robert Harding Picture Library; 62 Peter Parks/Oxford Scientific Films; 63 David Jeffrey/The Image Bank; 64/65 G. Noel-Figaro/Gamma/Frank Spooner Pictures; 66/67*t* & *b* The Natural History Museum, London; 68/69 François Gohier/Ardea; 71 Sinclair Stammers/Science Photo Library; 72/73 Brian Coope/Planet Earth Pictures; 74/75 Michael R. Schneps/The Image Bank; 76*l* Carol Farneti/ Oxford Scientific Films; 76*r* & 77*l* John Shaw/NHPA; 77*r* François Gohier/Ardea; 80/81 Joanna McCarthy/The Image Bank; 81 Hiroji Kubota/Magnum Photos; 82/83 © Westermann Schulbuchverlag GmbH; 84/85 David E. Rowley/Planet Earth Pictures; 86/87 CarolFarneti/Oxford Scientific Films; 87 Doug Perrine/Planet Earth Pictures; 88/89 Dr Georg Gerster/The John Hillelson Agency; 91 Stephen Dalton/NHPA; 92/93 Hiroji Kubota/Magnum Photos; 94/95 Dr Georg Gerster/The John Hillelson Agency; 96 François Gohier/Ardea; 96/97 John Shaw/ NHPA; 100/101 François Gohier/Ardea; 104/105 R. Ian Lloyd/Susan Griggs Agency; 105 John Cleare/Mountain Camera; 106 & 107 Harald Sund/The Image Bank; 110/111 Anthony Bannister/Oxford Scientific Films; 112/113 John Cleare/Mountain Camera; 115 John Shaw/ NHPA; 116/117 Larry Dale Gordon/The Image Bank; 118*l* Darodents/Zefa Picture Library; 118*r* Dr Robert Spicer/ Science Photo Library; 119*l* Bleibtreu/Sygma; 119*r* John Shaw/NHPA; 120/121 Doug Allan/Oxford Scientific Films; 122 O. Brown, R. Evans and M. Carle, University of Miami, Rosenstiel School of Marine and Atmospheric Science; 127 Nancy Sefton/Planet Earth Pictures; 130/131 Darodents/Zefa Picture Library; 132 Walter Looss/The Image Bank; 134/135 Earth Satellite Corporation/Science Photo Library; 137 Guido Alberto Rossi/The Image Bank; 142 University of Dundee; 142/143 Dr Robert Spicer/Science Photo Library; 145 F. Sauer/Zefa Picture Library; 147 Bruce Coleman; 151 NASA/Science Photo Library; 152/153 Brian Brake/The John Hillelson Agency; 154/155 NASA/ Science Photo Library; 157 Robert Harding Picture Library; 158/159 John Shaw/NHPA; 162/163 Gordon Garradd/Science Photo Library; 164 Dr G. T. Meaden/ Fortean Picture Library; 165 & 168 Robert Harding Picture Library; 169 Bleibtreu/Sygma; 170/171 Zefa Picture Library; 172*l* Steve McCurry/ Magnum Photos; 172*r* Herve Collart/Gamma/Frank Spooner Pictures; 173*l* NOAA/Science Photo Library; 173*r* Ron Sanford/Black Star/Colorific!; 174/175 Robert Harding Picture Library; 178/179 Herve Collart/Gamma/Frank Spooner Pictures; 179 Bartholomew/Gamma/Frank Spooner Pictures; 180 Guido Alberto Rossi/The Image Bank; 181 Michael Friedel/Rex Features; 182/183 Steve McCurry/Magnum Photos; 184 NOAA/Science Photo Library; 184/185 Zefa Picture Library; 187 NASA/Science Photo Library; 188/189 David W. Hamilton/The Image Bank; 192/193 Ron Sanford/Black Star/Colorific!; 196/197 NASA/ Science Photo Library.

Artwork credits
l = left; *r* = right; *t* = top; *c* = centre; *b* = bottom

Richard Bonson 24/25, 26, 70/71, 78/79, 88/89, 108/109, 114/115
John Davis 24/25
Gary Hincks 40/41*b*, 46, 90/91, 126, 132/133, 194/195
Gary Hincks/The Natural History Museum 66/67
Aziz Khan 12/13, 58, 94, 128/129, 134, 150, 152/153, 182
Mainline Design 18/19, 34/35, 38, 40/41*t*, 44, 48/49, 82, 98/99, 120, 122/123, 124/125, 136, 140, 160/161, 166/167
Janos Marffy 10/11, 60/61, 68, 84/85, 96, 106/107, 130, 142, 144, 146, 155, 158/159
Jonothan Potter 20/21, 36/37, 52/53, 102/103, 138/139, 141, 148/149, 156/157, 179, 186/187
Colin Salmon 22/23, 32/33, 65
Peter Sarson 72/73, 112/113, 176/177
Ann Winterbotham and Richard Draper 14/15, 16/17, 190/191
The graph on pages 98/99 first appeared in *New Scientist*, the weekly review of science and technology.

If the publishers have unwittingly infringed copyright in any illustration reproduced, they would pay an appropriate fee on being satisfied to the owner's title.